ASVAB STUDY GUIDE 2021–2022

COMPREHENSIVE REVIEW WITH PRACTICE TEST QUESTIONS FOR THE ARMED SERVICES VOCATIONAL APTITUDE BATTERY EXAM

TABLE OF CONTENTS

PART I: INTRODUCTION

one

THE BASICS

I f you have ever considered joining the military (or if you have ever spent time taking ROTC in either high school or college), then you have probably already heard about the ASVAB. ASVAB stands for ARMED SERVICES VOCATIONAL APTITUDE BATTERY, and its purpose is twofold. For one, your score is going to determine whether you can join the military at all and which branches of the military you are eligible to join (Marines, Army, Air Force, Navy, or Coast Guard). Beyond that, the test will help to determine which specialties you are eligible for once you have a branch selected.

This test is timed, and it tests multiple aptitudes. It is provided at all schools which have ROTC programs, and it is also provided at Military Entrance Processing Stations (MEPS). This means that the test is available at more than 14,000 locations across the United States. The test was developed and is primarily maintained by the United States Department of Defense. For individuals who do not live close to one of the 65 MEPS stations in the United States, the ASVAB can also be taken at Military Entrance Test (MET) sites. These are often located in National Guard armories, military reserve centers, or inside of office buildings run by the Federal government.

Who This Guide is Meant For

This guide is meant for any individuals who are thinking about joining the military and want to know more about the ASVAB and the process of taking it. The guide is also meant for anyone who is planning on taking the ASVAB and wants to brush up on their knowledge and learn exactly what they need to prepare for. One interesting thing to note, which many people do not consider, is that the ASVAB tests a broad level of general knowledge. Individuals might be tempted to either take the test or use this study guide simply to gauge their general knowledge levels.

The Scope of this Guide

This guide is going to take you through a good deal of information, including the following:

- what the test is about
- pre/during/post-test information
- individual test sections with study tips, information that will be covered in the test, and practice questions for each subtest
- generalized study tips which will help you moving forward, regardless of the test you are taking
- a comprehensive practice exam, close to what you will see when you take the ASVAB
- a section comprised of further reading which you can use to help bolster up your weak points

There is more to joining the military than just taking the test, of course, but this will give you a great place to start. The ASVAB is usually one of the very first things that you will do in the process of joining up, right after a cursory physical exam. Thus, this guide primarily covers information about the exam and what the exam covers.

How to Use This Guide

STOP! Before you read further, do the following (it will benefit you immensely): **GO AND TAKE A PRACTICE ASVAB EXAM AND RECORD YOUR SCORES.**

Now, why do you believe you should take that exam before moving any further? To set a benchmark, of course. If you have a starting point, you will know what aptitudes you are weak in prior to beginning your study. This will let you narrowly focus your study in exactly the areas you need.

Do you HAVE to take a practice exam prior to the test? Of course not. You can go through the guide straight away without taking any practice tests. But if you do that, you will miss out on valuable statistical information about the success of your study. Taking a practice test will also help you determine which aptitudes you need to spend more time on. Information is the key to success, so there is no reason to turn down the information a practice test can provide you (especially since it costs nothing).

Once you have completed the guide, go take another practice ASVAB and compare the resulting scores to the ones that you received before you began your study routine. Identify weak areas and repeat the process until you are satisfied with your scores. This is one of the best ways that you can find out how much you have improved and what you need to study more.

two

THE ASVAB

Taking the ASVAB, like it or not, is not like taking most standardized tests. For one thing, the number of different skills which are being tested is much higher. For another, it is more important than most standardized tests. If you blow it on the SAT, then you can still get into most colleges. It isn't going to affect your future in too big a way. If you blow the ASVAB, on the other hand, it could have a significant effect on your potential future. The ASVAB is what helps recruiters and officials determine your aptitude for certain occupations and specialties within the military and, in fact, what branches you are eligible to join (and whether you can even join at all). Take the test seriously, because it's important. Use this guide and any other study materials you have at your disposal in order to prepare for the test and get the most that you can from it. The temptation to skimp on your studying is there, just like with any standardized test, but you simply have to overcome that.

Pen and Paper vs. Computerized

There are two versions of the ASVAB, which are currently offered: A physical test (pen and paper) and a computerized version of the test. Both of these versions of the test offer unique challenges and, if possible, you should try to find out which one you will be taking ahead of time. That extra preparation will help give you the best possible chance of succeeding.

The Physical Test: Pen and Paper

The physical test is much like every other standardized test you have ever taken, but it most closely resembles the SAT or the GRE. Once everyone has arrived and has been seated in the testing room, the administrator will instruct everyone on how the test will be taken. They will then pass out sheets for answers and test booklets. This will include pencils if you have not brought your own (again – you do not need to bring one; they are available to you at the testing center).

The total time for the entire test is somewhere between three and four hours, usually closer to four. You should pay close attention to the instructions given to you by the test administrator and do not advance without being told to. All of the individual aptitudes being tested have their own subtest, which has both a time limit and a specific number of questions on it.

Once each subtest has been finished, you can go back and review the answers that you have chosen, assuming you have enough time left on that test. You are not able to review previous subtests that you have completed, however, and you cannot move on to the next subtest either. The time that you have is only permitted to be used in taking your current subtest or reviewing that subtests answers, and nothing else.

Here is a breakdown of the pen and pencil ASVAB:

Table 2.1. Pen and paper ASVAB content

SUBTEST	TIME LIMIT (MINUTES)	NUMBER OF QUESTIONS
General Science (GS)	11	25
Arithmetic Reasoning (AR)	36	30
Word Knowledge (WK)	11	35
Paragraph Comprehension (PC)	13	15
Mathematics Knowledge (MK)	24	25
Electronics Information (EI)	9	20
Auto and Shop Information (AS)	11	25
Mechanical Comprehension (MC)	19	25
Assembling Objects (AO)	15	25
Totals	**149**	**225**

The Computerized Test

The largest difference in the computerized version of the ASVAB (the CAT-ASVAB) is that it is taken on a computer, obviously. It is also adaptive. Each time you answer a question, correctly or incorrectly, the test will adapt to your ability level and modify the next questions. This allows the test to be taken in a shorter amount of time than you would have to take on the pencil and paper test.

Prior to beginning the test, an administrator is going to explain how the test is formatted and will give you a brief rundown on how to use a mouse and a keyboard. They will also explain how to answer questions, how to get help, etc. If you have questions, this is the best time to ask them.

Everyone takes the CAT-ASVAB at a different pace. When you have finished with one subtest, you can immediately move to the next subtest if you want. Time limits are imposed on the individual aptitude tests, but time will rarely expire before you have finished one. The amount of time and remaining questions for each test is displayed on the screen for you. It usually takes about an hour and a half to finish the CAT-ASVAB.

One difference in this and the pen and pencil version of the test that is important to note is that once you have submitted an answer, you cannot review or change it. Another change from the physical version of the test is that auto and shop information have been split into two separate aptitudes, rather than a single subtest.

Here is a breakdown of the computerized ASVAB:

Table 2.2. Computerized ASVAB content

SUBTEST	TIME LIMIT (MINUTES)	NUMBER OF QUESTIONS
General Science (GS)	8	16
Arithmetic Reasoning (AR)	39	16
Word Knowledge (WK)	8	16
Paragraph Comprehension (PC)	22	11
Mathematics Knowledge (MK)	20	16
Electronics Information (EI)	8	16
Auto Information (AI)	7	11
Shop Information (SI)	6	11
Mechanical Comprehension (MC)	20	16
Assembling Objects (AO)	16	16
Totals	**154**	**145**

What to Expect

Nothing can ever truly prepare you for the test like just taking it can. You can have all the knowledge in the world about what is on it and how it will go, but for 99.9 percent of the people taking it, it will be an entirely new experience in an unfamiliar place and, thus, pretty stressful. The information contained within this section should help ease your mind a bit by explaining some of the nuances involved in the taking of the test.

It should be noted that the version of the ASVAB that you take in high school if you are a member of ROTC, will be a bit different than the version you will take at MEPS. The purpose here is to provide potential career areas for high school students that guidance counselors can use. With that being said, you can still use these scores as your entrance scores for the military if you choose to go that route.

Pre-Exam Considerations

There are a number of things that are going to happen before you take the exam. First, you will need to see a military recruiter (specialized to the branch you are applying to join). The recruiter is going to screen you (and any other applicants) to make sure you will be a good fit and that your candidacy is valid. They will ask about if you have a criminal record, whether you are married, what sort of education you have, basic health-related information, and whether or not you have a history of drug use. These questions, obviously, should be answered as honestly as possible. They may go through a basic physical exam with you as well. Once they have qualified you to move on in the recruitment process, the next step is taking the ASVAB itself.

The test is either going to be administered at an MEPS station or at a METs station, depending on your location. At times, the recruiter that you have been working with will drive you to this station in order for you to take your test. They are not allowed to accompany you inside of the testing room itself, but they are permitted to take you there.

This should be obvious, but DO NOT BE LATE TO THE TEST. If you are late to the test, they will turn you away, and you will have to reschedule your test time. This is not ideal for anyone, so make sure you leave some bumper room in your schedule if you think you might end up cutting it close.

THE ONLY THING YOU NEED TO BRING TO THE TEST IS YOUR SOCIAL SECURITY NUMBER (SSN) AND A VALID PHOTO IDENTIFICATION. They will provide you with pencils and test booklets. Calculators are not permitted for use on the test. You may bring your own pencils if you choose, but you will not be required to have one and, again, they will be passed out by the test administrators on test day regardless.

What to Expect During the Exam

During the exam, you can expect a few things. Most of them might seem like common sense, but you should still be aware of them just in case. Here are some of the things you can expect.

If taking the physical copy of the exam, the times will be a bit longer. You will be given instructions by a proctor, and you will only be able to answer and review the current subtest that you are on. All test materials are provided, and all questions are asked of the proctor.

If you are taking the computerized version, you cannot go back and review your answers. The time limits are also shorter. This test adapts to your answers, so don't be surprised if the difficulty of the questions changes as you go along through the test.

The fact that there are time limits on the subtests might stress you out. Be prepared for this and don't let it get to you while you are taking the test. You will only be permitted to work on the current subtest you are on, regardless of the type of test you are taking. The ASVAB is only administered in English. The procedure is going to vary a bit depending on where you are taking the test, so it is never going to be exactly the same. If you are late to the test, you will have to leave and reschedule the day of your test. You won't be allowed into the testing room if you do not have a valid identification with you.

Post-Exam Considerations

Taking the exam is only part of the process. Once you are done, there are a few things you will still have to do. Get your scores, for one thing. Speaking with your recruiter and letting him or her know how the test went for you is another thing you may want to do.

After the exam, you can expect a few things:

- Your AFQT score will be calculated quickly by the test administrator and provided to your recruiter immediately regardless of which test you take.
- The physical test takes a few days to be scanned, and your score will usually be available afterward.
- If you take the computerized version of the test, your scores will be available to you immediately.
- Do not linger in the test room after the test. Once you are finished, turn it into the proctor and ask to be allowed to leave.
- DO NOT TALK ABOUT THE TEST once it is over. You will be subject to penalties if you write down any of the test questions or talk about the test once you are finished with it. Keep it to yourself and don't spread any information.

Exam Sections

Each section of the ASVAB is meant to test a specific aptitude. There are nine sections in total. Here are the sections covered on the ASVAB, with the possibility that auto/shop might be separated into two disparate sections, depending on whether you take the physical test or the computerized test:

1. General Science
2. Word Knowledge
3. Paragraph Comprehension
4. Mathematics Knowledge
5. Arithmetic Reasoning
6. Electronics Information
7. Auto and Shop Information
8. Mechanical Comprehension
9. Assembling Objects

One thing to note: The subtests are not necessarily going to be in the same order on each test. Do not put too much faith in them being in a specific order and take that into account when you are doing your studies.

Sections of the ASVAB

Here is a complete series of descriptions for each exam section:

Table. 2.3. Section expectations

APTITUDE	DESCRIPTION
General Science (GS)	The subtest is primarily concerned with the physical sciences and with principles of biology.
Word Knowledge (WK)	Meanings of words, antonyms, and synonyms, ect.
Paragraph Comprehension (PC)	Paragraphs of text followed by questions concerned with the content of what you have read.
Mathematics Knowledge (MK)	This is high-school level mathematics, which includes both geometry and algebra.
Arithmetic Reasoning (AR)	These are mathematical word problems requiring relatively simple math skills to solve. Logic based, usually.
Electronics Information (EI)	Principles of electronics, basic information about circuits, and terminology.
Auto and Shop Information (AS)	Information about automotive terminology, basic auto and shop skills, and the use of tools.
Mechanical Comprehension (MC)	Principles of basic mechnics and physics.
Assembling Objects (AO)	Orientation of objects in space (physical space—not outer space).

Scoring and Results

There are two different scores that you are going to get when you take the ASVAB. One of them will determine whether or not you are qualified to join the military and which branches you are able to join. The second is the most complete of the two scores. This is the score which is used to help determine your eligibility for various military occupations and specializations. The higher your score, the better your choice of potential specializations will be. It is not possible to get a perfect score, however, so the goal here is to do the best that you can do.

First, you will get what is known as the Armed Forces Qualification Test (AFQT) score. This is the score that will be used to determine whether you are qualified to join specific branches of the military (and whether you can join at all. The AFQT is the composite of the results of the arithmetic reasoning, math knowledge, and both verbal composite (word knowledge and paragraph comprehension) sections of the test.

Here is a list of service branches and the minimum associated AFQT score (these scores could change without notice, as stated in the military and Department of Defense guidelines):

Table 2.4. Minimum AFQT scores

SERVICE BRANCH	AFQT SCORE
Army	31
Navy	35
Marines	31
Air Force	36
Coast Guard	45

The AFQT score is, easily, the most important score of the two for most people. Doing well on this is what will determine whether or not you even have a future in the military. A good analogy would be to consider the AFQT the cake and the other subtests as the icing. If you don't have a good cake, you won't even need the icing.

It should also be noted that most special enlistment programs for individual branches are going to have their own minimum AFQT scores, which often differ significantly from the minimum scores required to join a specific branch:

- **Navy**: Requires anyone with a GED to have a minimum score of 50 for enlistment. Anyone wishing to get in on the college fund or the college repayment program needs also to have a score of 50.
- **Army**: A minimum score of 50 is needed for monetary enlistment bonuses, college repayment programs, and the Army College Fund.
- **Air Force**: A minimum score of 65 is required to enlist in the air force if you have a GED. Their programs are extremely selective, and they want highly skilled candidates.
- **Marine Corps**: The USMC requires a score of 50 on the AFQT for most programs. This includes their enlistment bonuses, the Marine Corps College Fund, and the Navy College Fund.

The second score that you receive will dictate which specializations you are able to enter. This is going to be the results from the other aptitudes which are included on the ASVAB. Unfortunately, the list of specializations that can be entered in the various branches of the military is far too broad and extensive to be listed in this guide. You can, however, find the scores you need and the full list of specializations on the websites for both the military (in general) and on individual branch sites.

If you take the pen and paper version of the test (the physical version), you will get your AFQB score immediately and you will have to wait a few days for the results of the other subtests. If you take the computerized version of the test, you will get your AFQB and your complete ASVAB scores immediately. No questions asked, no need to wait for results. They will be right there waiting for you.

When you get your scores back, you might be expecting one single, easy to understand, score. That is, unfortunately, not the case. You will get a whole bunch of scores. Interpreting these can, at times, be daunting for someone unfamiliar with what they are looking at.

Here are the types of scores you will get back.

- **STANDARD SCORE:** This is the method that your subtests will be reported back to you. They are based on a standard distribution that has a mean of 50 and usually has a standard deviation of 10. This is NOT a 1-100 grading system like you would have seen in school, so do not confuse it with that.
- **RAW SCORE:** This is the total of the number of points you have gotten on the subtests. Questions are weighted differently. These scores are used to help calculate the other scores, but raw scores will not usually be shown on your results card.
- **PERCENTILE SCORE:** These have a range of 1-99 and help to compare your scores with the "normal" group. It shows what percentile of test takers you are in. The higher your score here, the better you did, relatively speaking.
- **COMPOSITE SCORE:** Individually calculated scores for each individual branch of the military. They all have their own way of handling this.

The calculation of the AFQT score itself follows a relatively simple procedure. First the value of your Paragraph Comprehension score is added to your Word Knowledge score. Next, the resulting number will be converted to a scaled score (your verbal expression score) which has a range between 20 and 62. Next you will double that verbal expression score and add your Mathematics Knowledge and Arithmetic Reasoning scores to it. This will give you your raw score. Finally, that score is converted to a percentile score which compares the results of your test with other individuals who have taken the test.

Your percentile score is going to be used to help determine your trainability for each individual branch. The following table should help you to determine where you stand:

Table 2.5. Minimum AFQT scores

PERCENTILE SCORE	TRAINABILITY	CATEGORY
93 – 100	Oustanding	I
65 – 92	Excellent	II
50 – 64	Above average	III A
31 – 49	Average	III B
21 – 30	Below average	IV A
16 – 20	Markedly below average	IV B
10 – 15	Poor	IV C
0 – 9	Not trainable	V

Only 4 percent of applicants from category IV can be taken into the military. No applicants from category V may be taken. These figures come from the Department of Defense.

All ASVAB test scores are valid for a period of two years. If you do not qualify with your current test scores, you are allowed to retest, as dictated by each individual branch:

- **AIR FORCE**: No retesting is allowed for delayed entry programs. If you have no job preference but have qualified scores otherwise, you may retest. You can retest if the line scores of your subtests limit the matching ability of currently available jobs. You may retest to improve your scores, but the recruiting flight chief has to interview you in person to give approval prior to any retesting.

- **ARMY**: You may retest if your scores have expired, if you fail to meet the AFQT requirements, or if extenuating circumstances occur (having to leave the test due to an emergency, etc.).

- **MARINE CORPS**: Retests can be given once the previous test scores have expired and at the discretion of recruiters. They cannot be requested only on the basis that the scores were too low for enlistment programs.

- **NAVY**: If your scores have expired, the test can be retaken. The test can also be retaken if your scores are too low to enlist.

- **COAST GUARD**: Tests can be retaken after six months in order to raise scores to qualify for certain enlistment options. The recruiters may also authorize retests after 30 days at their discretion.

three

GENERAL TEST TIPS

E ven the simplest test will have some tips which are universal. In this section, tips are included to help you on the ASVAB in a generalized way, regardless of which type of test you are taking or which aptitude test you are currently on. In fact, most of these tips will help you on any kind of standardized test that might take, no matter what subject the test is on (MCAT, SAT, GRE, ASVAB, LSAT, etc.).

Here are some tips to utilize before the test. Begin preparing for your test as early as you possibly can. If you do not feel as though you are quite ready for it, then do not put pressure on yourself to take it early. With regard to the ASVAB, recruiters may, at times, try to railroad you into taking the test as quickly as possible (sometimes within a couple of days of speaking with them). If you feel like you are not ready, schedule the test for a later date.

Plan very carefully ahead of time and use your time wisely. Do not waste time or procrastinate on your studying. Study at the same general time every day. Don't let yourself become sidetracked doing other things when you should be spending the time studying. Make sure you study in a well-lit and quiet area. Somewhere comfortable. Avoid distractions if possible. This is a great time to utilize a public library. When you are taking practice tests, you will want to make sure that you time yourself in the same way that you will be timed on the real test. This will give you the greatest feel for how the test will actually be administered. Many people who neglect to do this wind up finding themselves short on time when the test is actually being taken.

Remember to get a good night sleep and eat well prior to the test. Remember to relax and do not allow yourself to become too stressed out about what is going on in your life. Don't eat a carb heavy meal the morning of the test, however, because it might make you tired during the test and affect your ability to concentrate. Avoid drinking lots of fluid (water or otherwise) prior to the test. You could wind up wasting a lot of the time during your test simply going to the bathroom. Don't make that mistake.

Make sure you bring supplies just in case. While you will be given pencils and paper, it would pay to bring pencils, erasers, pens, and scratch paper just in case. You can never be too careful. Make sure you have your contacts, a backup set, and glasses if you need them as well. A watch might help you keep track of time during the physical version of the test (in case there is not a clock in the room). It will pay to invest in a watch if you don't already have one just to be extra careful.

Here are a few tips for making the most of your study materials and practice sessions ahead of time. Make sure you study the information about the test just as much as you study what is on the test. This means the types of questions you might face, the format of the test, the length of time you have for each test section, etc.

Take adequate time to prepare by taking a practice test at regular intervals throughout your study. This will let you find your weak points so that you can narrowly focus on them. Besides the obvious, this will also alert you to weak points in areas that take longer amounts of time to study, such as vocabulary, paragraph comprehension, and grammar.

Do not spend too much time on your strong points. Review them, yes, but do not spend an inordinate amount of time going over materials that you already have a firm grasp on. Analyze why you missed questions on the practice test. Was it because you second guessed yourself? Was it because you just didn't know how to answer the question? Remember: the more information you have, the better.

Do not try to memorize the answers to questions. Figure out what the question is about and study it. Specific questions are not as important as having a firm grasp of the concepts being tested. Do not try to memorize everything in this guide (or any guide) in a small time window. It simply won't work. In fact, it will probably undermine the work that you are trying to accomplish.

Generalize the types of thinking that you will have to use during each aptitude test you will take. Make sure you spend time on every part of the guide during your studies. All of them are unique, and all of them will help. Multiple short study sessions are better than single long study sessions.

Here are some strategies you can use to help you narrow down answers. Break down the amount of time that you actually have to spend on individual questions. This can be found by dividing the test time by the number of questions. Make sure you read through every single answer prior to marking anything. You have to keep in mind that many tests want you to select the "best" answer, meaning more than one might be correct, but one of them is "more" correct than the others. Do not make any snap judgments about a particular answer without reading the others, because more than one might be technically correct.

Use clues within the question to help you select the answer and make an educated guess. Run through the test questions very quickly to get the easy ones out of the way first (for the pen and paper version, at least). That way you can determine which questions you will need to spend more time on. If all else fails, go with your gut. People will often second-guess themselves and end up marking the wrong answer. If you lean toward a certain answer right from the start, then go with it.

Eliminate any answers that you can right from the start. On multiple choice tests, two or sometimes three answers will often be obviously wrong. Cross through them or ignore them so that you do not waste valuable time. Always read the full directions before going to try and select answers. This is shockingly important, and it is something that many individuals, for whatever reason, do not do. Big mistake.

Finally, there is one other thing you need to remember: **FOLLOW THE RULES DURING THE TEST**. Don't do anything that might be mis-construed as cheating. Don't talk to anyone. Avoid shifting your eyes around to others. Only ask questions to the test administrator. Do not copy down any questions to your papers. If you finish a section early (on the physical version of the test), review your answers. Do not sneak looks at material not being covered yet.

PART II:
EXAM REVIEW

four

GENERAL SCIENCE

The purpose of this section of the ASVAB is to determine how well the applicant understands various (basic) concepts from the physical sciences, the life sciences, and Earth sciences. This is a section which is basically going to be all rote memorization on your part. That can be a daunting task, but it shouldn't be something which gives you pause. You will do just fine if you go through this material a couple of times. Most of it, in fact, will probably seem like common sense if you remember your basic science classes from the past.

One thing to keep in mind is that there are a lot of facts and figures which will show up in this section. You will see a ton of questions that cover everything from physics to chemistry. Rather than trying to memorize all of the little individual facts that you come across, it will pay for you to try and memorize the base foundations. This would be all of the general principles that are behind the little facts and figures. Consider it akin to looking at the "big picture" of the scientific world. That will send you on the way to a good score.

On the General Sciences subtest, you have eleven minutes for twenty-five questions. That is not a lot of time for each individual question, as you can probably gather immediately upon looking at those figures. The bright side, though, is that there is not a lot of interpretation that goes on in this section. Either you know the correct answer to a question or you do not know the correct answer to a question. It is as simple as that. Most of the questions that you will encounter here are going to take less than ten seconds to answer, and you have about twenty-six seconds on average.

For individuals who plan to (or already have) gone through everything in this entire study guide, you can relax a bit here. The General Sciences subtest is not one of the ones that count on your AFQT score. Does that mean that this section is not important? Nope. It doesn't mean that at all. It simply means that if you need to prioritize your studying, the four subtests that are on the AFQT should come first, this one later. If you are aiming for a specialty within the military that requires a high score on the General Science subtest, you will definitely want to go through this section with a fine-toothed comb.

Going through this section of the guide will give you a broad overview of what is on the subtest and will help you boost your skills. If you want to do well on this subtest, spend some extra time here. There are some additional resources that you can check out if you wish to bolster your skills a little bit more (these can be found at most university/college libraries and, if not, there will be comparable books at local libraries):

- *Biology* by Robert Brooker
- *Astronomy: A Beginner's Guide* by Erich Chaisson and Steve McMillan
- *General Chemistry* by Linus Pauling
- Any of the *For Dummies* books covering one of these topics will cover everything you need for the ASVAB as well.

It is important to note that nobody is allowed to know what exactly is asked on the ASVAB, so the best anyone can do is base their information on the practice tests and on a general level of knowledge.

Study Information

One of the first things you need to know, and something that will apply to every single subsection you will find below, is how measurement works in the scientific world. The system, which differs significantly from the system used in the United States, is known as the metric system.

Table 4.1. Metric system units of measure

PHYSICAL PROPERTY	BASIC UNIT	SYMBOL
time	second	s
volume	liter	L
mass	gram	g
length	meter	m

These units can be modified using prefixes based on the power of ten. For example, 1000 meters is equal to one kilometer.

Table 4.2 Metric system prefixes

PREFIX	ABBREVIATION	MULTIPLES
tera	T	10^{12}
giga	G	10^{9}
mega	M	10^{6}
kilo	k	10^{3}
hecto	h	10^{2}
deka	da	10
deci	d	10^{-1}
centi	c	10^{-2}
milli	m	10^{-3}
micro	u	10^{-6}
nano	n	10^{-9}
pico	p	10^{-12}
femto	f	10^{-16}
atto	a	10^{-18}

If you are wondering how these two tables fit together, think about this example.

What is a ks? We know from the first table that *k* means kilo and that kilo means 1000 (10^3). From the second table, we know that *s* means second, the basic metric unit of time. So *ks* is a kilosecond or, 1000 seconds.

So what about temperature? Most people know there are at least two temperature scales which are commonly used. The Celsius [C] scale and the Fahrenheit [F] scale. There is also a third, the Kelvin scale. This scale comes into play during high-level chemistry and physics discussions. Now the three scales will be broken down, and conversions between C and F will be discussed.

◆ KELVIN: This scale is a bit more specific than the other two. It is used to help determine what the absolute coldest temperature possible is. On this scale, absolute zero is defined as the temperature at which molecular motion would cease to occur. That temperature is 0 K. For a bit of reference, the freezing point of water would be 273.15 K.

◆ FAHRENHEIT: The standard temperature scale in the United States. Water freezes at 32°F and water boils at 212°F.

◆ CELSIUS: The metric standard scale for temperature across the world. The freezing point of water is 0°C, and the boiling point is 100°C.

There are two types of conversion systems. One is a bit more complex than the other.

Table 4.3. Conversion systems

	SYSTEM ONE	SYSTEM TWO
From °F to °C	$C = \frac{5}{2} \times (F - 32)$	$C = [\frac{5}{2} \times (40 + F)] - 40$
From °C to °F	$F = \frac{2}{5} \times (C + 32)$	$F = [\frac{2}{5} \times (40 + C)] - 40$

This may seem complicated, but you have to bear in mind that the two resulting equations are the exact opposite of each other insofar as the multiplication is concerned, so if you know one of them you can easily figure out the other.

The Scientific Method

One of the things that sets the modern sciences apart from the methods done in the past is a system of hypothesis testing known as the scientific method. Through this, new ideas can be cultivated and tested in order to help form the basis for the various sciences.

Here are the basic steps of the scientific method:

1. OBSERVE things going on around you or in the universe in general.
2. FORM A HYPOTHESIS (an educated guess) about why your observation occurs.
3. PREDICT AN OUTCOME based on that hypothesis.
4. EXPERIMENT using the hypothesis and prediction as a basis and record your results. If they do not match, modify the hypothesis.
5. REPEAT STEPS 3 AND 4 until you are able to come up with a repeatable result.

Disciplines

A number of scientific disciplines are covered on the ASVAB.

Table 4.4. Scientific disciplines

SPECIALTY	DESCRIPTION
agriculture	the study of farming
archeology	the study of past anthropology. Civilizations, tools, etc.
astronomy	study of space
biology	the study of life
botany	the study of plant life
chemistry	the study of chemicals, elements, and reactions
ecology	the study of the environment
entomology	the study of insects

genealogy	the study of ancestry
genetics	the study of genes and heredity
geology	the study of rocks and minerals
ichthyology	the study of fish, a sub-specialty of biology
meteorology	the study of weather
paleontology	study of prehistoric life

All of these individual disciplines are covered by the subheadings below. Knowing the niche specialties can be important, however, so don't skimp here. With that being said, the questions won't be as simple as "what does a meteorologist do".

Life Sciences

Life sciences cover exactly what you might think that it covers. Biology and systems related to it. This includes everything from the smallest microbe to the largest organism and everything in between. It also includes the classification systems that we use to help piece together the bigger picture of biology.

Classifications

Classifications are how scientists denote differences and relationships between different organisms. Every single organism is going to have its own designation (usually two words, the first is the GENUS and the second is the SPECIES). Most of these names are derived from Latin, so a basic familiarity with that language will help a lot here (and with nearly all of the other sciences as well).

The following is a list of the classification system (all eight levels):

1. domain
2. kingdom
3. phylum
4. class
5. order
6. family
7. genus
8. species

To see a full example of this in action, let's look at humans:

Table 9.5. Human classification

domain	eukaryota
kingdom	animalia
phylum	chordata
class	mammalia
order	primates
family	hominidae
genus	homo
species	homo sapien

As you can see, the classification gets more and more specific the closer to the species it gets. Finally, you arrive at the smallest distinction that you can get to at the species level.

As far as kingdoms are concerned, most scientists agree that there are five:

- **FUNGI** are multi-celled organisms that do not undergo photosynthesis. Mushrooms belong to this kingdom.
- **PROTISTS** are single-celled organisms which have a nucleus.
- **MONERANS AND VIRUSES** are single-celled, non-nucleated, organisms. This includes bacteria, algae, and viruses. Viruses are sometimes argued to be their own kingdom, as they are unique in the world of biology.
- **PLANTS** is one of the two largest kingdoms (animals is the other). This includes non-moving organisms with no obvious sensory or nervous systems. They have cell walls consisting of cellulose.
- **ANIMALS** are multi-celled organisms without cellulose-comprised cell walls, chlorophyll, or the ability to undergo photosynthesis. Organisms in this kingdom can both respond to external stimuli and can move.

Evolution

EVOLUTION is a term which describes a change in characteristics of a population of organisms over a given period of time. The process is usually slow and gradual. One important thing to note is that evolution does not have an end goal. It is a process which is constantly ongoing and constantly changing. There is no beginning or end to the process.

These principles can be used to help sum up the way evolution works:

- A huge amount of genetic variation exists among different organisms of the world.
- Organisms have to compete for a limited supply of resources.
- The organisms which can survive the best and are able to reproduce will be naturally selected for continuation because of that.

Two important things to note with regard to natural selection: genetic variation is random among individuals, and traits allowing an organism to survive and reproduce are going to be passed down to the offspring of that organism. Traits that do the opposite will die off. Organisms which are adapted better to their particular environment are going to be more likely to survive and, thus, their traits will be more likely to live on in subsequent generations.

Some factors which can influence and contribute to the process of evolution include the following:

- GENETIC DRIFT: When groups leave larger populations and establish new populations, they can become genetically different from their parent populations, leading to speciation (at times).
- MIGRATION: This occurs when individuals move into and out of a given population, allowing gene flow between multiple populations.
- MUTATION: The process of genetic material replicating itself is not perfect. In fact, there are systems built into the process which cause mutations in order to potentially adapt organisms better to their environments. These can be harmful or beneficial to the individual.

So how do scientists know that evolution is real? How do they know this is something which actually happens? Ever since the publication of *On The Origin of Species* by Charles Darwin, scientists have been collecting evidence and striving to prove the theory. One of the largest places from which they derive their evidence is through the fossil record, which shows the way modern organisms have descended from common ancestors (think birds from dinosaurs).

Another common way this is done is through comparative anatomy. Many different types of organisms have similar structures. The bones of the arms of a human, the wings of a bird, or the fins of a porpoise, for example, all share some common bones and traits, though the exact shape may differ amongst them.

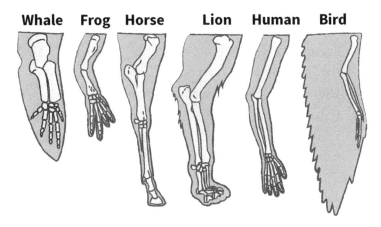

Figure 4.1. Homologous structures

The above image shows some common structures across different organisms.

Embryology is another area which is commonly studied when looking into evolution. The embryos of many animals look nearly the same until they begin to differentiate, suggesting that they all come from the same ancestor organism. Chicken embryos and human embryos are a particularly striking example of this.

Finally, the study of genetics and biochemistry has greatly enhanced the ability of scientists to look at evolutionary and genetic changes on a very small scale (at the DNA and RNA level).

The Cell

Cells are the most basic unit of every living organism. Some organisms are, in fact, just single cells. Others, like the complex organisms that humans, animals, and most plants are, are comprised of multiple cells which are both specialized and organized into groups that make up tissues and organs. Every single cell is going to have two basic things: a PLASMA MEMBRANE and CYTOPLASM. The plasma membrane is the outermost boundary of the cell. This is what keeps the inside of the cell inside and keeps the outside of the cell outside. Cytoplasm is the liquid-like substance which makes the basic foundation of the cell. This is located inside of the plasma membrane and helps to give the cell a shape.

Cells are either PROKARYOTIC or EUKARYOTIC. Prokaryotic cells are usually very simple, like bacterial cells. These are almost universally going to be single celled organisms and will not comprise any substantial parts of multicellular organisms. Eukaryotic cells are more complex and contain specialized organelles within them which carry out special functions.

The primary components of eukaryotic cells are NUCLEI, MITOCHONDRIA, RIBOSOMES, and CHLOROPLASTS:

- The NUCLEUS is where genetic information is stored and accessed within the cell.
- The MITOCHONDRIA is where ATP is produced, the primary energy molecule in most cells. This is colloquially known as the "powerhouse" of the cell.
- RIBOSOMES are where cells make proteins.
- CHLOROPLASTS are found only in plant cells; this is the site of photosynthesis. This is where plant cells make their energy and convert it into sugar for storage.

It is worth noting that there are many other organelles in most eukaryotic cells. These, however, are the most basic and most important ones. These are the ones you are most likely to encounter on the ASVAB.

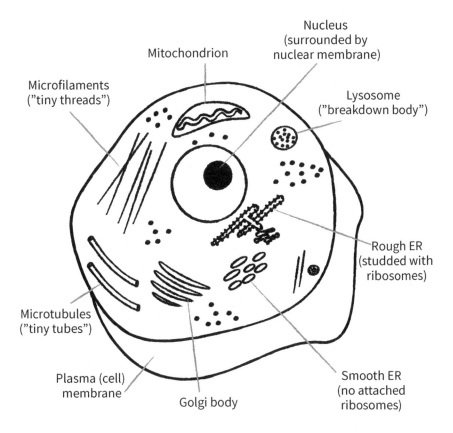

Labels on diagram:
- Microfilaments ("tiny threads")
- Mitochondrion
- Nucleus (surrounded by nuclear membrane)
- Lysosome ("breakdown body")
- Rough ER (studded with ribosomes)
- Microtubules ("tiny tubes")
- Plasma (cell) membrane
- Golgi body
- Smooth ER (no attached ribosomes)

Figure 4.2. Eukaryotic cell

The plasma membrane, which you will remember is the outermost boundary of the cell, is studded with proteins and other mechanisms which help with the selective permeability which is required for substances to move in and out of the cell. There are four primary ways through which things can pass inside and outside of plasma (cell) membranes:

1. **ACTIVE TRANSPORT**: Molecules are able to cross the plasma membrane going from regions of low concentration to regions of high concentration using specialized proteins located within the membrane itself. This process requires the expenditure of energy reserves (ATP).

2. **FACILITATED DIFFUSION**: Specialized proteins allow diffusion across the cell membrane during the right conditions.

3. **OSMOSIS**: This is a specialized form of diffusion which concerns itself with the movement of water across the cell membrane, usually to equalize pressure or change relative concentration.

4. **DIFFUSION**: This is the passive movement of molecules from regions of high concentration to regions of low concentration (think about putting a drop of food coloring into a glass of water).

Respiration and Fermentation

The way that many organisms receive energy is known as the process of cellular respiration. During this process, cells break down carbohydrate molecules, usually glucose through a process known as glycolysis (*glyco* means sugar, *lysis* means to break apart). The energy produced through cellular respiration is stored as adenosine triphosphate (ATP). When cells require energy, they break down the ATP and then utilize the energy from the breaking of the bond. Respiration requires the presence of oxygen. It takes place within the mitochondria.

The process of cellular respiration can be summarized as:

$$\text{glucose (sugar)} + O_2 \text{ (oxygen)} > \text{water} + CO_2 \text{ (carbon dioxide)} + \text{ATP (energy)}$$

When oxygen is not available, cells of a few organisms will switch to a process known as anaerobic (*an* means without; *aerobic* means oxygen) respiration. This process is also known as fermentation. The products of this process are ethanol and carbon dioxide.

Photosynthesis

Photosynthesis is the process through which plants make the food that they use for energy. They use sunlight, carbon dioxide, and water in the process of photosynthesis to make glucose. The process requires energy, which is where the sunlight comes into play. Sunlight is captured in specialized organelles (chloroplasts) which use pigments (chlorophyll) to capture different wavelengths of light.

The process of photosynthesis can be summarized as:

$$CO_2 \text{ (carbon dioxide)} + \text{water} > \text{glucose (sugar)} + O_2 \text{ (oxygen)} + \text{water}$$

Interestingly enough, plants are green because they do not use green light and, instead, they reflect it back.

Cell Division

The cells of living organisms have the ability to reproduce exact copies of themselves through division. This, along with the expansion of cells, is how organisms grow and develop.

MEIOSIS is the process through which an organism creates four daughter cells from the parent cell. Each of these will have half of the number of chromosomes that the parent cell has. These daughter cells are gametes (eggs and sperm) which are used for sexual reproduction.

MITOSIS is the process most cells go through. Two copies of the original cell are created. There are a number of phases to this process. Each of the two daughter cells will have the same number of chromosomes as the parent cell has.

Genetics

Genetics is an enormous subject. It is growing larger every single day, and the body of knowledge is absolutely staggering. With that being said, only the basics of genetics are going to be covered on the .

Genetics is the study of genes and how they relate to the traits and function of organisms. Genes themselves are small parts of DNA molecules which relate to specific functions and traits in the larger individual to which they belong. Each gene can have more than one form. These individual forms are known as alleles. Which allele an organism inherits from parent organisms is how traits manifest themselves.

Here are some common terms which you should know for the ASVAB:

- TRANSCRIPTION is the process through which DNA is copied onto RNA.
- TRANSLATION is the process through which RNA is translated into proteins.
- GENOTYPE is the sum total of the genes in an individual.
- PHENOTYPE is the way those genes express themselves in characteristics of an individual.
- The RECESSIVE ALLELE is the allele which is not expressed when they are paired with a dominant allele. Geneticists usually use a lower case letter to represent the alleles which are recessive.
- If two different alleles are present, and one of them is expressed, then that is the DOMINANT ALLELE of the two. This is usually going to be shown as a capital letter.
- HETEROZYGOTE is an individual who has two different alleles for the same gene.
- HOMOZYGOTE is an individual who has two identical copies of the same gene.

The CENTRAL DOGMA of molecular biology is that the flow of information flows from DNA > RNA > protein.

Note: These definitions are very simplistic. They will suit you for the , but you may want to brush up on them individually in order to get a better understanding of the core concepts.

Human Anatomy and Reproduction

The study of anatomy and physiology should start with diet and nutrition. This is something which is often overlooked in many textbooks. In addition, it is a somewhat confusing topic because of the "buzzword" nature of the whole thing. Everyone knows about fad diets and the latest food trends. Not many people, however, know much about the basics of what foods are composed of.

Here is a brief rundown of nutrients obtained from food sources:

- **VITAMINS** are rganic compounds which are essential to health. Generally, these are going to be proteins. Some break down easily in fats, and some break down easily in water.
- **MINERALS** are elements required by animals to build new proteins and to function properly. Think phosphorous, magnesium, calcium, and zinc (among many others, of course).
- **LIPIDS** form cell membranes, and they help to comprise some hormones. Fat is used as a storage source for energy as well.
- **NUCLEIC ACIDS** are used to form RNA and DNA, nucleic acids are gained from eating other nucleic acids (and then subsequently breaking them down).
- **CARBOHYDRATES** are the basic source of energy all animals use. Glucose is the most commonly used carbohydrate for energy, and it is created during the process of cellular respiration.
- **PROTEINS** form the framework and most of the structures found in the body. They are comprised of amino acids which are gained by eating other protein sources.

The Digestive System

The human digestive system is composed of a number of parts which work together in conjunction. An important aspect of the digestive system is the assistance of a variety of gut bacteria in the process of digestion.

Here are some of the major components of the human digestive system and their function:

- The **MOUTH** is a familiar organ to most people. This is where you break large food particles into smaller ones through a process known as mastication (chewing). Saliva also helps to begin the digestion process.
- The **ESOPHAGUS** is a highly muscular tube which carries food from the mouth down to the stomach. Food is moved through a rhythmic contraction of the muscles located there (the process known as peristalsis).
- The **STOMACH** is an expandable bag-like organ inside the abdomen. It is filled with gastric juices (and acids) which break down food. The food is then moved to the small intestine.
- The **SMALL INTESTINE** is where absorption of food occurs, primarily.

- The **LARGE INTESTINE** is where remaining waste is elimi-nated and turned into feces which can then be passed out through the anus to leave the body.

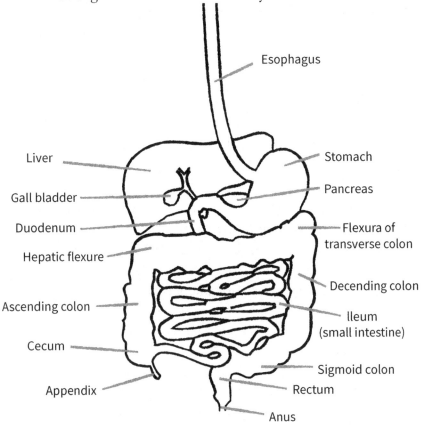

Esophagus
Liver
Stomach
Gall bladder
Pancreas
Duodenum
Flexura of transverse colon
Hepatic flexure
Decending colon
Ascending colon
Ileum (small intestine)
Cecum
Sigmoid colon
Appendix
Rectum
Anus

Figure 4.3. Digestive system

The Respiratory System

The respiratory system consists primarily of the following organs:

- The **LUNGS** consist of small sacs called alveoli; this is where capillaries from the circulatory system undergo gas exchange, releasing carbon dioxide, and binding with oxygen.
- The **TRACHEA** is where air branches into the two lungs.
- The **PHARYNX** passes air from the nose into the throat.
- The **NOSE** is where the respiratory system begins. Air is moistened and otherwise conditioned here. Hairs in the nose help to trap dust and to purify the air.

The Excretory System

This is the system that gets rid of the waste products that the body makes. The human kidneys are two organs located near the stomach and the liver. This is where blood is filtered. Once it has been filtered, waste products are carried through ureters (tubes) to the urinary bladder where it is stored until it can be released. Urine is composed of uric acid and urea, among other things, which are the waste products resulting from the breakdown of nucleic acids and amino acids, respectively.

The Nervous System

The human nervous system is comprised of two sub-systems. The central nervous system and the peripheral nervous system.

The CENTRAL NERVOUS SYSTEM consists of the brain and the spinal cord. Of the two, this is the most important. The central nervous system is where sensory information is processed and where your consciousness, intelligence, and memories are stored.

Extending from the base of the brain down to the base of the spine, the SPINAL CORD connects the peripheral nervous system to the brain. Impulses and signals travel up and down the spinal cord, which acts as a center of coordination for the nervous system as a whole.

- The cerebrum, thalamus, hypothalamus, and limbic system make up the FOREBRAIN. The forebrain controls sensory input, learning, memory, emotion, and the synthesis of hormones.
- Between the fore- and hindbrains, the MIDBRAIN consists primarily of nerves and fibers used to alert the forebrain should something out of the ordinary occur.
- The medulla, cerebellum, and pons are what make up the HINDBRAIN. The hindbrain helps to bridge the regions of the brain and to help coordinate muscle movements.

The PERIPHERAL NERVOUS SYSTEM is all of the other nerves in the human body. These are the nerves that connect the central nervous system to all of your peripheral parts.

- The AUTONOMIC NERVOUS SYSTEM is comprised of the sympathetic and the parasympathetic nervous systems. This is an involuntary series of nerves which help to prepare the body for emergency situations (fight or flight, for example) and helps with sleeping and preparation for digestion.
- The sensory somatic system is the system which carries any impulses from the environment and the senses, which allows people to be both aware of their environment and to act on the information they are receiving.

The Circulatory System

The human circulatory system is the system which transports oxygen and other nutrients to the tissues of the body from the lungs and the intestines.

Here are some of the most important components of the circulatory system:

- **Blood** is comprised of plasma, platelets, red blood cells, and white blood cells.
- Red blood cells are also known as **erythrocytes**. These are donut shaped cells with no nucleus. They carry **hemoglobin**, which gives them their red color, and binds oxygen and carbon dioxide to carry it to and from body tissues.
- White blood cells are known as **leukocytes**. These are a major part of the human immune system.
- **Platelets** are small and disc-shaped fragments of cells which are created in the bone marrow. These are the starting material that is used to initiate the blood clotting process.
- **Plasma** is a straw yellow colored fluid like substance comprised of other blood proteins and constituents, including waste products, hormones, and nutrients. Its primary constituent is water.
- **Lymph** is a watery fluid which is derived from blood plasma and is an active component of the immune system.
- The **heart** is the organ which keeps your blood flowing through your body. It does this through a rhythmic series of contractions which occur involuntarily.

Ecology

Ecology is a specialty of biology which concerns itself with interactions between the environment and organisms in that environment. Here are some basic definitions you will need to know in order to navigate this section of the :

- An **ecosystem** includes every species and every organism which is living in a certain community and how they interact with each other. This includes things such as soil, water, and light as well.
- A **population** is members of a single species which are living in a pre-determined area of the environment.
- A **community** is the specific living area of animal and plant species.

There are many ways that organisms in a specific ecosystem may interact with one another. **Predation** is when one one organism feeds on another one. One of the organisms in the situation will benefit while the other is injured in some way. **Parasitism** is when one organism benefits and the other is harmed. In parasitism, one organism will live off of or feed on another, usually without killing it outright. **Commensalism** is when one organism benefits while the other is not affected one way or another. Lastly, **mutualism** is when both organisms benefit from this type of relationship.

Here is a basic layout of the ways that energy flows through a given ecosystem, being transferred from one place to another:

- **PRODUCERS** are the organisms which trap the energy of the sun through photosynthesis (e.g., plants and algae).
- **PRIMARY CONSUMERS** are the herbivores which feed on the producers of the environment.
- **SECONDARY CONSUMERS** are usually carnivores. These are the organisms which feed on the primary consumers of the environment.
- **DECOMPOSERS** are the organisms which help to break down dead organisms and then recycle their nutrients into the environment.

The study of ecology is important, but it usually takes a backseat to more formal studies of the individual subject areas of life science. With that being said, ecology is becoming more and more prevalent as a field of study given the rise in environmental awareness all around the world.

Chemistry

Chemistry is one of the physical sciences and is one of the basic sciences that form the general science subtest of the . For this subtest, you will be concerned with the composition and structure of matter, including the properties of matter and the way that matter changes over time and due to environmental factors. The primary concern of chemistry is with atoms and molecules, both the way that they interact with one another and the transformations that they can undergo.

Chemistry is arguably the most critical science. Nearly every other scientific discipline needs to utilize information from chemistry in order to move forward with their specific research. It also helps that chemistry is the bridge between physics, biology, and the other sciences. It is a type of physical science, but it should not be confused with physics.

Atomic Structure the Nature of Matter

So if chemistry is basically at matter, what, exactly *is* matter? **MATTER** is anything that has a mass and that takes up a given volume. The three most common states of matter are all things which you are familiar with already: gasses, liquids, and solids.

- **SOLIDS** have a defined shape, volume, and mass.
- **LIQUIDS** have a defined volume and a defined mass, but they do not have a definite shape.
- **GASSES** have a definite mass but no volume or shape. They expand to fill whatever they are put in.

The state of matter that a particular element or molecule is found in is determined by the amount of kinetic energy it has. **KINETIC ENERGY** is the energy of motion. All particles of matter are moving. They are constantly moving. The speed at which they are moving is what primarily determines their state of matter. Gasses move the fastest, and then liquids, then solids.

Temperature can play a role in this. When you heat up a solid, it will become a liquid because the particles speed up. When you heat a liquid, it becomes a gas. When you apply cold, the opposite happens. Heat speeds up the motion of molecules, while cold slows down the motion of molecules.

There is another form of matter which bears mentioning as well: plasma. PLASMA is an ionized state of matter which is somewhat similar to a gas. Don't expect to see many questions about plasma on the —what questions you do see will most likely not be anything complex.

Atoms, Molecules, and Compounds

The basic building block of everything in chemistry is the atom. The atom is generally defined as the smallest unit of matter which defines a chemical element. Every type of matter is made of up one or more atoms, either neutral in charge or ionized (with a charge). These are very small particles, measured in picometers.

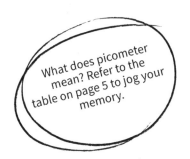

What does picometer mean? Refer to the table on page 5 to jog your memory.

Atoms are composed of a few subatomic (smaller than atoms) building blocks. The first of these is the nucleus, which is the core of the atom. Electronics are negatively charged particles which circle around (outside of) the nucleus. Inside of the nucleus are protons (positive charge) and neutrons (neutral charge). The vast majority of the mass of an atom is within the nucleus. It should also be noted that most atoms are going to be of neutral charge. There will be as many electrons as there are protons, so they will cancel out each other's charge. Additionally, there are usually as many neutrons as there are protons. If the number of neutrons differs from the number of protons, then that atom is known as an ISOTOPE. You may be familiar with this concept through terms like CARBON 13 or something similar. Isotopes are frequently used to track molecules and atoms moving through systems. Below is a table outlining the most common subatomic particles and some of their characteristics.

Table 4.6. Subatomic particles

PARTICLE	CHARGE	MASS IN GRAMS	SYMBOL
electron	−1	9.109×10^{-28}	e
neutron	0	1.675×10^{-24}	n
proton	+1	1.673×10^{-24}	p

For example, hydrogen, the simplest atom, contants 1(+1) proton, 0 neutrons, and 1(-1) electrons.

The electrons that are located outside of the nucleus form a "cloud" or a number of SHELLS (the most common way of describing them, and essential to advanced forms of chemistry).

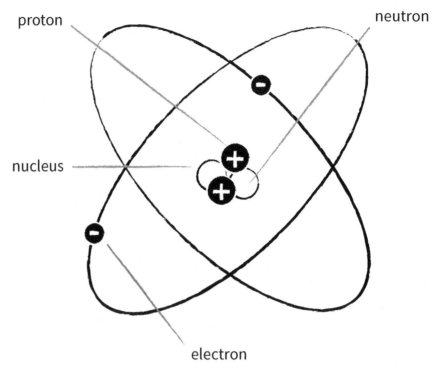

Figure 4.4. Structure of an atom

Given all of the different combinations of protons and electrons, there is an almost infinite number of atoms which could be formed. At the current time, however, only 122 are known. There are others that have been theorized, and some which have been created artificially, but only 122 are known to exist.

The next thing that you need to concern yourself with is molecules. You must know by now that everything is not made up of just individual atoms. Different combinations of atoms go together to form molecules and compounds. Compounds are substances which have more than one type of atom bound together with it. Molecules are compounds which are formed of tightly bound nonmetals (this will make more sense after the discussion of the periodic table). That is a distinction which can be somewhat confusing, but it important to remember.

There are a number of elements which are always found in nature as molecules. These are known as DIATOMIC MOLECULES. Some of these elements are only found in nature either in compounds with other elements or bound to another element of the same type. Here are the most common diatomic molecules in nature:

◆ oxygen (O_2)
◆ fluorine (F_2)
◆ hydrogen (H_2)
◆ nitrogen (N_2)

Compounds have properties which differ from the individual elements from which they are composed. This is evident, readily, in the case of, say, hydrogen (H_2) versus water (H_2O). At room temperature, hydrogen is going to be a gas. At room temperature, water is going to be a liquid. Obviously not the same, but both of them nevertheless have hydrogen as a constituent part.

The Periodic Table

The periodic table has been designed in order to easily display elements, along with their most common characteristics, in one easy to read table. In this table, elements are assigned either a one-letter or a two-letter designation which is known as their atomic symbol. For example hydrogen is represented by the letter H, helium is represented by the letters He, and so on. The first letter is always going to be capitalized and the second letter is always going to be lower case. This is a method of keeping elements straight when they are written out with other elements in chemical formulas (more on this later).

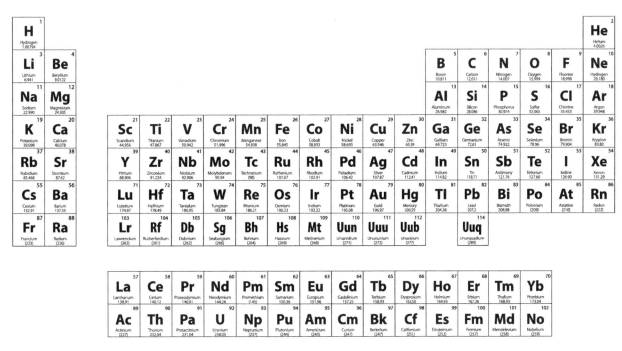

Figure 4.6. The periodic table

At first glance, this table may appear to be confusing or nonsensical. Once you understand the method by which it was designed, however, you will quickly discover that it is able to hold an enormous amount of information in a relatively small space.

The periodic table is read from left to right, and the atoms on the table have been arranged by the number of protons they have within their nucleus. This is also known as the ATOMIC NUMBER. Hydrogen has one proton, so it comes first on the table. It should be noted that at a certain point, the number of protons between atoms on the table is no longer a difference of one. For example, the difference in Ra and Lr is

88 and 103, and they are found right beside each other on the table. The atomic number (the number of protons) is the characteristic that defines an atom of a particular element. All atoms of the same element have to have the same number of protons. They do not, however, have to have the same number of neutrons or electrons.

If an atom has the same amount of protons and electrons, then the atom is going to be neutral. The positive and negative charges of that atom will cancel each other out. If there are fewer electrons than there are protons, then the atom will be a positively charged CATION. In the opposite case, an atom with more electrons than protons is going to be a negatively charged ANION. Isotopes are neutrally charged atoms with a different number of protons than neutrons, which has already been discussed.

The ATOMIC WEIGHT (atomic mass) of an element is going to be listed on the periodic table right beneath the symbol. The atomic mass is going to be the average mass of all of the isotopes of an element which can naturally occur. Additionally, the columns of the table indicate the group number for individual elements. The groups indicate how many electrons are in the outermost shell of an atom. These outermost electrons are commonly called VALENCE ELECTRONS. They are the electrons which hold different atoms together when they form molecules.

The table itself was given its name because of the trends in the elements that appear when they are arranged in order of atomic number. This shows differences between elements which are regular (periodic) but are not readily apparent. This includes delineations between nonmetals and metals in addition to other things (semimetals, halogens, etc.).

Reactions and Equations

Studying elements and molecules is great, but how is it going to help you understand the chemical processes which are going on all the time? The answer to that question is in the study of reactions and chemical equations. Chemical reactions and equations are the shorthand which chemists use to describe chemical processes.

On the following page is an example of a basic chemical equation which can be used to illustrate the basic concepts:

$$C + O_2 > CO_2$$

This reaction is a COMBUSTION REACTION which sees carbon and oxygen combine to form carbon dioxide. The two reactants, oxygen and carbon, of the equation are found on the left-hand side of the arrow. The arrow itself is what represents the reaction occurring. The right-hand side of the equation shows the product(s) of the reaction. In this case, that product is carbon dioxide. The LAW OF CONSERVATION OF MASS states that matter can be neither created nor destroyed, only modified. In this situation, it means that you need to have the same elements and the same number of atoms on both sides of the reaction. As you can see in this example, there is one carbon atom on both sides and two oxygen atoms on both sides. The equation has been balanced.

There are a few types of chemical reactions that you can expect to encounter on this subtest of the . A DECOMPOSITION REACTION is a reaction in which a substance breaks down into multiple constituent components (at least two). Below is the formula for this type of reaction, as well as an example of a common decomposition reaction:

$$AB > A + B$$

$$2H_2O > 2H_2 + O_2$$

A SYNTHESIS REACTION (combination reaction) is a reaction in which at least two substances react with each other to form a single compound. Below is the formula for this type of reaction, as well as an example of a common synthesis reaction:

$$A + B > AB$$

$$3H_2 + N_2 > 2NH_3$$

A SINGLE REPLACEMENT/DISPLACEMENT REACTION is a reaction in which a single element reacts with a compound and exchanges places with one of the component elements of that compound. Below is the formula for this type of reaction, as well as an example of a common single replacement/displacement reaction:

$$AB + C > AC + B$$

$$3AgNO_3 + Al > Al(NO_3)_3 + 3Ag$$

A DOUBLE REPLACEMENT/DISPLACEMENT REACTION is a reaction in which two compounds react in order to achieve an exchange. Below is the formula for this type of reaction, as well as an example of a common double replacement/displacement reaction:

$$AB + CD > AD + CB$$

$$AgNO_3 + NaCl > AgCl + NaNO_3$$

In the above reactions, notice that the same elements in the same quantities are found on the product side and the reaction side. Mass is neither created nor destroyed. All of the equations have been balanced.

Both types of replacement/displacement reactions are collectively known as substitution reactions.

Solutions, Bases, Acids

Solutions, bases, and acids are some of the most common types of chemicals you will encounter in your day to day life. The following are some definitions to help get you started:

◆ The pH SCALE measures how much acid is in a solution (the power of hydrogen in the solution). A pH between 0 and 7 is acidic. A pH of exactly 7 is neutral. A pH between 7 and 14 is basic.

- An ACID is any compound that raises the amount of hydrogen (H+) ions in a solution.
- A BASE is any compound that decreases the H+ concentration in a solution by increasing the concentration of hydroxide (OH–).
- A NEUTRAL SOLUTION is a solution which is neither acidic nor basic. pH of 7 is neutral.
- A SOLUTION is a homogeneous mixture which is composed of two things: a solvent and a solute.
- A SOLVENT is the material which is in larger proportion within the solution (the one doing the dissolving).
- A SOLUTE is the material which is being dissolved in the solvent.
- A BUFFER is a substance which helps to ease the change in pH when acids or bases are added to a solution.

If you are a bit confused as to the definitions between a solution, solvent, and solute, take the example of salt water. Salt water is a solution in which the water acts as the solvent and the salt (sodium chloride [NaCl]) acts as the solute.

Water, in fact, is known as the universal solvent because of how many different substances it is capable of dissolving. Water is neutral (when pure). Below are some of the most common bases you have likely encountered:

- SODIUM CARBONATE, Na_2CO_3, is used in the manufacturing of paper and is also used to "soften" water.
- LIME, CaO, is typically used to raise the pH of the soil for growing certain kinds of crops. It is a commonly used substance in the farming industry.
- AMMONIA, NH_3, is a chemical which is used commonly in household cleaners. It is also used in fertilizers. It is one of the primary constituents of urine.
- LYE (sodium hydroxide), $NaOH$, is a commonly used base which is used in order to make soap.
- ACETONE, C_3H_6O, is commonly used as a solvent. It is often used for stripping paint.

Below are some common examples of acids:

- SULFURIC ACID, H_2SO_4, is the chemical compound with the highest amount of industrial production in the entire world. It is also utilized in the creation and operation of car batteries.
- PHOSPHORIC ACID, H_3PO_4, is commonly used in soft drinks in order to help stem potential bacterial growth that could

occur in them. It is also used in the production of many commercial fertilizers.

- **VINEGAR** (acetic acid), HC_2H3O_2, is a solution of acetic acid (usually 5 percent acetic acid in water).

- **CITRIC ACID**, $H_3C_6H_5O_7$, is found in fruits. It is what gives them their slightly acidic and tangy taste. In addition, it is utilized in the metabolism of many animals

- **CARBONIC ACID**, H_2CO_3. This acid is the acid which everyone talks about being in carbonated soft drinks. The acid is formed when carbon dioxide is dissolved in water.

- **NITRIC ACID**, HNO_3, is a type of acid that is commonly used in the creation of fertilizers (due to the nitrogen content).

- **HYDROCHLORIC ACID**, HCl, is a very common acid. You may also know it as stomach acid.

The Most Important Elements

Of all of the elements, the first twenty which are listed in the periodic table are the most important. They are the most abundant on Earth, and they are likely the most abundant in the entire universe. The twenty elements listed below are used in materials that people use every single day, in both industries and in their own bodies. The format for these elements is as follows: name (symbol/atomic number): information.

1. **HYDROGEN (H/1)**: Hydrogen is the most abundant element in the universe. It is clear, odorless, and low density. It is also highly flammable, which is why it is no longer commonly used for blimps and balloons. It is found in nature as a diatomic element (H_2). Hydrogen ions which have been dissolved in water cause that water to become acidic.

2. **HELIUM (HE/2)**: Helium is a clear gas with no odor and low density. It is also the second most abundant element found in the universe. It is not, however, one of the most abundant elements found on Earth. Helium is not flammable, so it is commonly utilized in lighter-than-air applications such as blimps and balloons. Helium is not particularly reactive, and it does not occur as a diatomic molecule when found in elemental form.

3. **LITHIUM (LI/3)**: Lithium is a metal with a low density. It is reactive when it is found on its own. It forms a +1 ion very easily. One of the most common uses of elemental lithium is in the treatment of some psychological disorders (particularly bipolar disorders).

4. **BERYLLIUM (BE/4)**: Beryllium is a metal with a low density. It is commonly used in technology due to the

strength that it imparts. At the same time, it is very toxic to humans, so care must be taken when it is handled.

5. **Boron (B/5)**: Boron is a gas at room temperature. It is classified as a metalloid, which has some of the properties of metals. When boron is turned into an oxide, it is used to make some cleaning agents and is a major constituent of heat resistant glassware.

6. **Carbon (C/6)**: Carbon may be the most important element that there is, right after hydrogen. Carbon is a solid when it is found at room temperature. It is also an extremely versatile element. Graphite, the substance used to make pencil lead, and diamonds are both forms of carbon. Carbon is capable of forming four bonds with other elements (or with itself). Carbon is also the basis of nearly all known life forms and is the major distinguishing element between organic and inorganic chemistry. Carbon dioxide is one of the major products of fossil fuel burning and, as such, is one of the commonly cited greenhouse gasses that environmentalists and policy makers concern themselves with.

7. **Nitrogen (N/7)**: Nitrogen, in its natural form, is an odorless and clear gas with no color. Three-quarters of the atmosphere of the Earth is composed of nitrogen (more than oxygen, in fact). This is one of the diatomic elements (it appears in nature as N_2). Nitrogen is commonly found in nitrates and in ammonia which are both utilized in the creation of fertilizers. Plants and animals are unable to metabolize nitrogen by themselves and, instead, rely on "nitrogen-fixing" bacteria to metabolize the nitrogen for their use.

8. **Oxygen (O/8)**: Oxygen is a flammable, odorless, colorless gas which reacts readily with nearly every element in order to form "oxides" (molecules with oxygen). This is another diatomic molecule, appearing in nature as O_2. Oxygen makes up around one-fifth of the atmosphere of the Earth. The ozone layer of the Earth is comprised of O_3 molecules.

9. **Fluorine (F/9)**: Fluorine is usually found in nature as an extremely reactive pale yellow gas. It occurs as a diatomic molecule (F_2). Fluoride is an anionic form of fluorine which is used in making toothpaste and helps to strengthen tooth enamel.

10. **Neon (Ne/10)**: Neon is found in nature as a nonflammable, nonreactive, odorless and clear gas in nature. It is commonly utilized in the creation of neon

signs, but there are often other gasses which are also used for that purpose as well.

11. **SODIUM (NA/11)**: When found in its elemental state, sodium is a soft, yet solid, shiny metal that is very reactive. When mixed with water, for instance, it is able to break the bonds between the hydrogen and oxygen in the water molecule with enough heat and force that the hydrogen will combust. Sodium forms +1 cations easily, and it is usually found in this state when it is found in nature. Sodium is one of the most common elements you will find, usually bonded with others. For instance: table salt is NaCl, a compound comprised of sodium and chloride.

12. **MAGNESIUM (MG/12)**: This is a shiny and solid metal which reacts with water when found at room temperature. When ignited, magnesium burns brightly. It is often used in match heads, some striker strips for ignition, and in other products used for combustion. It is also a major constituent part of the chlorophyll molecules which are used by plants during the process of photosynthesis.

13. **ALUMINUM (AL/13)**: This is a lightweight and shiny metal, solid at room temperature, which is used in a variety of applications. It is used in everything from soda cans to the siding of houses. It forms an oxide coating which can prevent further reactions, which is partially what makes it so useful. It is used in areas where rust would be a concern. Before a standard process was found to isolate elemental aluminum, it was one of the rarest metals. So rare, in fact, that as a sign of the dominance of the United States, the top of the Washington Monument is capped with aluminum.

14. **SILICON (SI/14)**: Silicon is a solid semimetal which is used to create a variety of objects, including quartz, glass, and sand, among other things. One of the largest modern applications for elemental silicon is the creation of computer chips. It is also utilized in many building applications, including caulking, where it is turned into a polymer known as silicone.

15. **PHOSPHOROUS (P/15)**: This is a solid element at room temperature and occurs in three primary colors: black, red, and white. Red and white phosphorous are the most commonly found forms and are usually found as P_4 when they are found in nature. This is an extremely reactive element and usually has to be stored under special conditions (underwater, typically) so that it does not react with the oxygen found in the air and ignite. Fertilizers often use phosphorous compounds.

16. **Sulfur (S/16):** This is a yellow solid at room temperature. It is very brittle. When found in nature, it is found in the form S_8. It is a major constituent part of sulfuric acid, and its primary industrial application is in the creation of that acid. Sulfur emissions from industrial burning (such as coal power plants) can result in the eventual formation of H_2SO_4 (sulfuric acid) in the atmosphere. When this is brought down with rain, it is referred to as *acid rain.*

17. **Chlorine (Cl/17):** Chlorine is an element everyone is familiar with. Elementally, it is a yellow gas with a very choking odor. It is diatomic in nature (Cl_2). It forms anions, −1, very readily and, as a result, it is often utilized in substitution reactions. This is a major part of the brine which is found in the ocean. It is also used as a way to sterilize surfaces, pools, and other things. When released into the atmosphere, it contributes to the destruction of the ozone layer.

18. **Argon (Ar/18):** A clear and odorless gas, argon is nonreactive. It makes up a very small portion (1 percent) of the atmosphere. It is usually used in light bulbs, where it helps to cover filaments. It is also used in some industrial applications, such as welding when there is a need to avoid reactions.

19. **Potassium (K/19):** This is a soft and solid metal which readily reacts when it is found in its elemental state. Potassium can react with water in a more violent fashion than even sodium is capable of doing. Only the +1 cationic form of potassium is found in nature. This is an extremely important element, playing a role in fertilizers which assist in plant growth and in the sodium/potassium pumps which help to assist in muscle contraction.

20. **Calcium (Ca/20):** Calcium is a metal which reacts with water in a way similar to the way magnesium does. It is not a strong enough reaction to cause an ignition, however. Calcium is usually found as a cation (+2) in nature. It is the major element making up both skeletal structures and tooth enamel. It is also used in a number of basic solutions, such as lime and calcium bicarbonate.

One thing to keep in mind when you look at the list and, in fact, when you look at the periodic table in general, is the fact that a given element and its symbol are not always obviously connected. You can look at potassium, for example, which starts with a P and has K as its symbol.

When you look at this list, you will probably realize that there are some properties which are unique to specific groups (columns). This is another benefit of the periodic table: substances which have similar properties will be grouped together. Below is a table of the most important groups on the periodic table of the elements.

Table 4.7 . Significant periodic table groups

GROUP	ELEMENTS	CHARACTERISTICS
1A; alkali metals	Li, Na, K, Rb, Cs, Fr	form salts that can be dissolved in water; readily react to make +1 ions
2A; alkaline earth metals	Be, Mg, Ca, Sr, Ba, Ra	readily react to make 2+ ions
5A; pnictogens	N, P, As, Sb, Bi	readily react to form −3 ions
6A; chalcogens	O, S, Se, Te, Po	readily react to form −2 ions
7A; halogens	F, Cl, Br, I, At	readily react to form −1 ions
8A; noble gasses	He, Ne, Ar, Kr, Xe, Rn	none are particularly reactive; do not form compounds or ions regularly; outermost electron shells are already filled

Out of the twenty important elements that have already been outlined, six of them are even more important. Those elements are carbon, hydrogen, nitrogen, phosphorous, sulfur, and oxygen (CHNOPS is a common acronym for these six elements). These elements are the most important for life.

Organic Chemistry

Organic chemistry is one of the largest and most widely studied branches of chemistry. It is the primary bridge between biology and chemistry as well. Organic chemistry is the study of molecules based on carbon. Carbon is able to attach to many other carbon atoms in order to form longer, chained, molecules, so the variety of possible molecules is extremely large. Some examples of things studied by organic chemists are nucleic acids, amino acids, oils, tissues, proteins, plastics, and a slew of others (far too many to name).

The simplest carbon-based molecules (organic compounds) are named based on the number of carbon atoms that are found in a single chain. Below is a table of the most common prefixes used in organic chemistry.

GO ON

Table 4.8. Common prefixes

PREFIX	CARBON ATOMS IN CHAIN
meth-	1
eth-	2
prop-	3
but-	4
pent-	5
hex-	6
hept-	7
oct-	8
non-	9
dec-	10

Organic compounds are what forms the basis of all life on Earth. They are also the primary driver behind the vast majority of experiments conducted by chemists. The way that carbon is able to bond, forming four total bonds, allows for such a wide array of diverse carbon-based compounds that all the different proteins, pharmaceuticals, and types of life on Earth are able to stem from them.

Metals

Most elements found on the periodic table are metals. Many metals are found in nature in an oxidized state (combined with either sulfur or oxygen). Some metals, like the ones commonly found in mines (gold, silver, copper, etc.) can be found in an elemental state, however. Metals are capable of forming ALLOYS. Alloys are mixtures of two or more metals. Allows are solid. You may be familiar with the term AMALGAM if you have been to the dentist a few times. This is a mixture of mercury and another metal. A given amalgam can be either liquid or solid. The state is dependent on the amount of mercury which is present in the mixture.

Metals all share a few common properties, all of which are outlined below:

- They are usually silver in color.
- They are shiny.
- They readily conduct both electricity and heat.
- They can be cut into thin sheets (sectile).
- They can be drawn into thin wires (ductile).
- They can be hammered into thin sheets (malleable).
- They are solid at room temperature.

Metals will usually have all of these properties at differing levels, however some don't follow all of the rules. Mercury, for example, is liquid at room temperature, not solid.

Energy and Radioactivity

Energy is a property which objects can have that can be transferred to other objects or can be converted into another form. In the same way that the law of conservation of mass states that mass cannot be created or destroyed, the law of conservation of energy states that energy cannot be created or destroyed. It can, however, be converted into different forms. Energy can be present as a reactant in a reaction or as the product. It can also be utilized to make the reaction happen as well.

Two types of energy exist: potential and kinetic energy. POTENTIAL ENERGY is energy that is being stored. It has the "potential" to be used and to do work. This is usually dependent on either the types of chemical bonds that are present or, in a more real world example, the distance an object is from the ground. A rollercoaster at the top of its track is going to have a lot of potential energy, for example.

KINETIC ENERGY is the type of energy being used for motion. If something is moving slowly, it has a low kinetic energy. The faster that object starts to move, the higher its kinetic energy is going to go. This type of energy is often shown in temperature. When things are hot, their atoms and molecules are moving faster than when they are cold.

The energy which is being stored inside the nucleus of a given atom is a form of potential energy. This is the type of energy that is being used in radiation treatments, in nuclear power plants, and in the creation of nuclear bombs. When a nucleus which is unstable converts (through decomposition) to a stable nucleus, that potential energy is released.

The following are a few definitions which can help you get through any questions about radiation:

- RADIOACTIVITY occurs when particles are released from a nucleus as a result of instability in the nucleus.
- GAMMA RAYS are high energy particles of light.
- An ALPHA PARTICLE is two neutrons and two protons.
- A BETA PARTICLE is an electron.
- A NEUTRON is a neutrally charged subatomic particle with a weight similar to that of a proton.

Physics

Physics is the study of matter and the motion of matter through space and time (surely you can see how this subject overlaps a bit with chemistry). Some concepts related to this include energy and force. Physics is broadly concerned with the natural laws which govern the way that the universe as a whole is going to behave.

Motion

Motion is what happens when something moves from one place to another place. Three types of motion exist: translational motion, rotational motion, and vibrational motion.

Put bluntly, TRANSLATIONAL MOTION is motion happening in a straight line (on a linear axis). It is defined by a change in the position of an object over a given period of time and by movement relative to a fixed reference point. ROTATIONAL MOTION is a motion that is happening around an axis, and lastly VIBRATIONAL MOTION is motion which is occurring around a single, fixed, point.

Physicists use a variety of terms in order to quantitatively define motion, including speed, velocity, acceleration, scalar, and vector.

Generally, SPEED is going to be a measure of how quickly something is moving by dividing the distance it has traveled by the time is has taken to do so:

$$\text{speed} = \frac{\text{distance}}{\text{time}}$$

VELOCITY is also a way to measure how quickly something is moving. However, as opposed to speed, velocity needs a third piece of information: direction. Velocity is known as a vector quantity (more on this later).

$$\text{velocity} = \frac{\text{displacement}}{\text{time}}$$

ACCELERATION is a change in velocity over a given period of time. When that velocity increases, acceleration is the term which is used. When it decreases, the term deceleration is used. This is also a vector quantity. It is positive when it is occurring, in the same way, a given object is moving and negative when it is moving the opposite way.

$$\text{acceleration} = \frac{\text{change in velocity}}{\text{time}}$$

A SCALAR quantity is one in which direction does not make a difference. A VECTOR quantity is one where the direction traveled does make a difference.

Graphs are used in order to quickly show the way an object is moving. Generally, two types of graphs are going to be used. The first is a POSITION-TIME GRAPH. This shows how the position of an object is changing over time. The velocity on this type of graph is going to be equal to the slope of the line on the graph.

The second type of graph is a VELOCITY-TIME GRAPH. This type of graph is meant to quickly show how the velocity of a given object is changing over a period of time. The slow on this type of graph is going to be the acceleration. This can also be used to figure out the distance being traveled by an object using the area underneath the line on the graph.

LINEAR MOTION is motion which is occurring in a single dimension, along the x-coordinates of a graph. Motion on the y-coordinate is known as **MOTION IN A VERTICAL PLANE**. Usually, the acceleration along either of these is going to be constant. Since that is the case, the acceleration itself can be considered a constant. During this situation, motion can be described by the **EQUATIONS OF KINEMATICS**.

KINEMATICS: IMPORTANT EQUATIONS

These equations are the main equations which are used in order to calculate acceleration and velocity. To get started, you need to understand the five primary variables utilized in these equations:

- x = displacement (distance traveled)
- a = acceleration
- t = time
- v = final velocity
- v_0 = initial velocity

The relationships between these five variables gives you the following four equations:

$$v = v_0 + at$$
$$x = \left(\frac{1}{2}\right)(v_0 + v) \times t$$
$$x = v_0 t + \left(\frac{1}{2}\right) \times at^2$$
$$v^2 = v_0^2 + 2ax$$

These equations all have four out of the five variables that were defined at the outset. That is the method through which you can find a missing variable. Which four are on the question you are looking at? Pick the right equation for that situation, plug in the relevant numbers, then solve for the variable you need.

Now that these definitions and equations have been outlined, it is time to talk about motion which is occurring in a vertical plane. One thing to note at the outset: The force of gravity causes an acceleration (on Earth) of 9.81 m/s^2. When you use the kinematic equations to describe an object moving in freefall, you will utilize g (gravity) instead of acceleration, in the same way, you will use y instead of x.

THE LAWS OF MOTION: ISAAC NEWTON

Isaac Newton was one of the most profound physicists ever to live. His ideas and findings are still used to this day in order to help describe motion. Before outlining his three laws of motion, the term force needs to be defined. A **FORCE** is a push or pull which can cause a resting object to move or cause a change in the velocity of an object that is already in motion. The three laws of motion (in order) are as follows:

1. An object at rest will stay at rest, and an object that is in motion will remain in motion at constant velocity unless a force acts upon it.

2. The acceleration of an object is directly proportional to the force that is acting on it and is inversely proportional to the mass of the object.

3. If an object exerts a force on another object, the second object will exert an equal force back on the first object, but in the opposite direction.

Newton's first law describes a concept known as INERTIA and is commonly known as the law of inertia. Inertia is the tendency of a given object to resist any change to its motion.

The second law of motion is meant to describe how forces act on objects. The general equation that is derived from the second law is the following, where m = mass and F = force:

$$F = ma$$

Force is another vector quantity. It has a size and a direction. Like velocity, is can be positive or negative, depending on the direction it is being applied.

Energy

Energy and work are terms which go hand in hand. When scientists talk about what mass a given object has, they are also referring to the amount of energy it has. Energy has two main types, kinetic and potential energy, as discussed previously.

Both types of energy will change when an object is either doing work or when work is being done to that object. **WORK** can be defined as a transfer of energy to an object when that object is put into motion because of a force acting on it. The following formula can be used to define work, where F = force (newtons), W = work (joules), and d = distance (meters):

$$W = Fd$$

So what is energy, anyway? Energy is the capacity to do work. If you pick up a ball and lift it over your head, you are doing work and supplying that ball with potential energy. The force that you are working against is the force of gravity, in that case. The work that is being done against the force of gravity is known as GRAVITATIONAL POTENTIAL ENERGY. The following equation can be used in order to calculate the gravitational potential energy of an object, where PE = potential energy (joules), m = mass (kilograms), g = acceleration due to gravity, and h = height:

$$PE = mgh$$

When an object is at a height, it has potential energy. When it begins to fall, however, it is going to be accelerated by the force of gravity and, thus, lose some of the gravitational potential energy that it has. Because of the law of conservation of energy (as discussed previously), energy is simply being converted from potential energy to kinetic energy.

The decrease in one type of energy is immediately associated with an increase in the other type. Kinetic energy of moving objects also has an equation associated with it, where KE = kinetic energy (joules), m = mass (kilograms), and v = velocity (meters per second):

$$KE = \left(\frac{1}{2}\right)mv^2$$

Many types of objects utilize the above principles in order to change potential energy to kinetic energy and vice versa (and types of energy between one another). Here are a few of them:

- **SOLAR CELLS** convert light energy into electrical energy.
- **BATTERIES** convert chemical energy into electrical energy.
- **LAMPS** convert electrical energy into heat and light.
- **HEATERS** convert electrical energy into heat.
- **FIRE** converts chemical energy into heat and light.

There are, of course, many others. Dams, generators, car motors, alternators, etc. All of these things utilize one form of energy in order to do a conversion. Many of them are essential to the way people live in the modern age.

There is one other term which needs to be discussed here: power. **POWER** is the rate in which something is able to convert energy from kinetic energy to potential energy (or the other way around) using work. Here is the equation used to determine power, where P = power (watts), W = work (joules), and t = time (seconds):

$$P = \frac{w}{t}$$

Fluid Mechanics

Fluid mechanics is the study of fluids and the way that they react to physical pressures. A **FLUID** is a substance which is going to offer minimal resistance to changes in the shape that it has when pressure is applied. Both liquids and gasses can be considered fluids. Solids, however, are not. Many properties help to define fluids as a whole, but the most important is the capability of fluids to exert pressure on things.

What is pressure? Pressure is a force which is being exerted per unit of area. The general equation for defining pressure, where P = pressure (pascals), F = force (newtons), and A = area (square meters), is as follows:

$$P = \frac{F}{A}$$

The way that fluids are able to exert pressure on things is explained by something known as the **KINETIC MOLECULAR THEORY**. This theory states that particles making up various fluids are continuously undergoing random motion. The particles will, as a result, constantly be colliding both with each other and with surfaces that they make contact with. When these particles make that contact, a force is exerted. The combined force of all of those collisions is pressure.

There are a three primary principles which govern the way that fluids work. The first one of these principles is **ARCHIMEDES' PRINCIPLE**. This principle states that if an object is covered by a fluid, it is going to be buoyed up by a force that is equal to the weight of the fluid that is being displaced by that object. The following formula can be used to determine the size of that force, where F = force (newtons), ρ = density (fluid), V = volume (fluid), and g = acceleration (gravity):

$$F = \rho V g$$

Any object which is being immersed in a given fluid is going to either float or sink. Which one it does depends on the weight of the object being immersed relative to the force that is being exerted on it by the fluid it is immersed in. If the force is less than the weight of the object, then the object is going to sink. If the two values are equal, the object can float in the liquid at any given depth. If the force is greater than the weight, the object will float on the surface of the liquid. This principle is frequently used to determine the weight of an object by measuring its fluid displacement in water and, in fact, that was the reason that it was originally created.

The second principle is Bernoulli's Principle, and the third is Pascal's Principle.

Electricity

Electricity is the phenomenon which is associated with the presence of an electrical charge. A wide range of things are associated with electricity including electric current, lightning, static electricity, and electromagnetism. Electricity is primarily caused by a flow of electrons.

The following are a few definitions that you should get to know:

- **ELECTRICITY** is the flow of electrical energy from one place to another (usually from an electrical power source to whatever is using it).
- **ELECTRIC CURRENT** is the method by which electrical energy is transported; it is the flow of electrons. Current is measured in amperes.
- The **LOAD** is the component or part of a given circuit which is consuming the electrical power in an electrical system.
- An **ELECTRON** is a negatively charged subatomic particle.
- A **CIRCUIT** is a system of metal wires capable of conducting electricity which then provide a pathway for current to flow.
- A **CONDUCTOR** is any substance that allows the flow of an electric current.
- An **INSULATOR** is any substance which does not allow for the flow of an electric current.

- **VOLTAGE** is the potential difference between two ends of a conductor. This is what allows for the flow of the electric current. Voltage is measured in volts.
- **RESISTANCE** is the strength to which a material "resists" the flow of electrical current. Resistance is measured in ohms.
- **RESISTIVITY** is a property of materials which helps to determine the amount of resistance that those materials have to the flow of electrical current.
- **ELECTROMAGNETISM** is the property of moving electrical charges to produce a magnetic field. Likewise, magnetic fields which are changing are able to generate currents of electricity.

Some of these terms are going to go together. The higher the voltage, the higher the current. The lower the voltage, the lower the current. Insulators and conductors are often used in conjunction with one another in order to create methods of carrying electrical current over long distances without losing any of the energy. Think about the rubber (insulator) wrapped around the copper (conductor) wires in your house, for instance.

Ohm's law is the equation and theory which govern the way voltage, resistance, and current relate to each other. It states that the current between two points in a conductor is going to be proportional to the potential difference (voltage) between those points, where I = current (in amperes), V = voltage (in volts), and R = resistance (in ohms):

$$I = \frac{V}{R}$$

You can use this equation in order to solve for any one of the three variables inside of it, provided you know the other two. This is a powerful equation which is used in many fields, including one that is covered in another the electronics subtest.

Sound

Sound is a type of vibration which moves through a given medium as a mechanically derived wave of both displacement and pressure. The medium through which sound waves travel could be air, water, or another fluid-like substance. The sound which is perceived changes depending on the type of medium through which the waves are traveling.

Sound waves are pressure variations which are being carried through matter. There are two types of pressure variations: RAREFACTIONS, areas which have low pressure, and COMPRESSIONS, areas which have high pressure.

These two pressure variations are created when the sound wave makes the molecules of air (or whatever medium they are traveling through) collide with each other and, thus, make the pressure variations move away from the sound source. One of the logical conclusions that you may be able to draw from this (or perhaps you already knew) is that sound cannot travel in a vacuum. Because vacuums inherently have no particles to be moved, the sound waves and pressure variations cannot propagate within them.

How quickly sound is able to travel depends on temperature as well. For example: in the air, at room temperature, sound waves can travel about 343 meters per second. Another variable which can affect the speed at which sound travels is pressure.

It is worth noting that sound can travel through mediums besides air as well. It can travel through solids and liquids. The speed at which sound travels through solids and liquids, however, is faster than it is through the air. When sound waves hit solid surfaces, they can be reflected back. You know these reflections as ECHOES. How long an echo takes to travel back is going to depend on how close the surface it is reflecting off of is to the original source of the sound wave. Some animals use echoes in order to figure out where they are, spatially, and to help locate things. Bats do this, as do dolphins. This is also the method through which sonar works on marine vehicles.

The combination of compressions and rarefactions is also called the wavelength. The number of waves coming in a single second is known as the FREQUENCY of the sound. The term PITCH is sometimes used in order to describe the same thing. If the source making the sound is moving, then someone listening will hear the sound changing into different frequencies as a result. When moving away, the frequency goes down, when moving closer, the frequency goes up. This trend of changing frequencies is what scientists call the DOPPLER EFFECT and is commonly used in applications such as ultrasound, sonar systems, and radar detectors.

Earth Science and Astronomy

Earth science is a term which encompasses all of the various fields of science that deal with the Earth itself. This is one of the oldest historical sciences. Included in it are the studies of the stars (astronomy), the oceans (oceanography), the Earth itself (geology), and the weather (meteorology). These are not, of course, the only subjects which are studied in Earth Science. Everything from the atmosphere and the hydrosphere to the biosphere is all included in this catch-all term. Further, other general science principles are utilized in the study of Earth science, including mathematics, physics, biology, and chemistry. The environment and ecology are also a primary focus of some scientists studying Earth science.

On the ASVAB, you will encounter questions in astronomy, oceanography, geology, and meteorology.

Astronomy

Astronomy is a science which is concerned with the study of celestial objects and everything that occurs outside of the Earth's atmosphere. This includes everything from planets and stars to, at times, the nature of the universe itself (though that is usually left to a branch of science known as physical cosmology).

Below is a brief list of some vocabulary you might encounter in the astronomy portion of the general science subtest:

- A PLANET is an astronomical object which is large enough to have its own gravity but not large enough to undergo thermonuclear fusion. It is also large enough to have cleared the local area of other celestial objects smaller than it.

- A STAR is a luminous plasma ball which is being held together by nothing but its own gravity. The Sun is a star, as are the vast majority of lights seen in the night sky.

- A SOLAR SYSTEM is a star and all of the objects which are orbiting that star. The objects do not necessarily have to be planets.

- A GALAXY is a system of stars, interstellar gas, dust, dark matter, and stellar remnants which are bound together by an immense gravity field.

- The SUN is the star found at the center of the solar system. It is also the primary source of energy for the planets which orbit it (including the Earth).

- An ELLIPTICAL PATH (or elliptical orbit) is a type of orbit around an object which is roughly egg-shaped, rather than being perfectly spherical.

- HELIOCENTRIC is a term that describes a system which has a sun as its center, such as our solar system.

- A ROTATION is the circular movement of a given object around a single point (usually the center of the object).

- REVOLUTION might be considered another term for orbit, astronomically speaking. This term is used when one object moves around another one.

Here are some facts about the Earth:

- The rotation and revolution of the Earth, along with its tilt, are what lead to the day and night cycle.

- The axial tilt toward the sun along with the revolution are what lead to the years and seasons.

- Local weather is usually a result of the amount of solar radiation which is able to enter the atmosphere.

- The axis of rotation of the Earth is 23.5 degrees.

- The Earth rotates along its axis while, at the same time, moving counterclockwise around the sun.

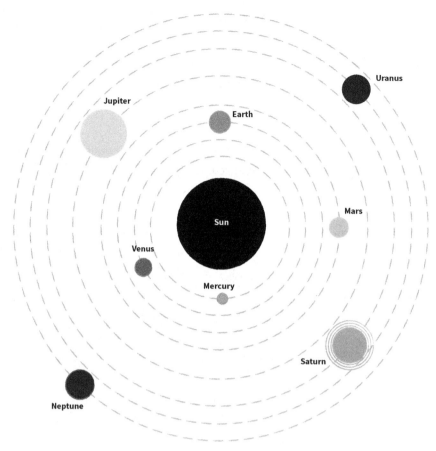

Figure 4.6. The solar system

The sun of the solar system that the Earth is in puts out such an immense gravitational field that it is able to hold all of the planets in their orbits around it. That solar system is also heliocentric, meaning that the sun is the center of the solar system. There are eight planets in the solar system which follow elliptical orbits around the sun. The output of electromagnetic radiation from the sun is what provides the energy from which life on Earth is derived.

MERCURY is the planet which is closest to the sun. The size of this planet is not much larger than the moon of the Earth. On the day side, the temperatures reach around 840°F while, on the night side, temperatures can drop to very far below freezing. There is nearly no atmosphere on mercury and, thus, there is nothing to protect it from impacts from meteors.

VENUS is the second planet from the sun. It is even hotter than Mercury, and the atmosphere is toxic due to a greenhouse effect (which also traps heat). The pressure on the surface of the planet is immense. Venus has a very slow rotation around its axis and spins in the opposite direction than most of the planets of the solar system. Venus is one of the brightest objects in the sky, outside of the moon and the sun.

EARTH is the third planet from the sun. This is a water dominant planet, having around two-thirds of the surface covered by water. This is the only known planet which has life on it. The atmosphere consists primarily of nitrogen and oxygen. The term *day* and *year* are based on the rotation of the Earth around its axis and the revolution of the Earth around the sun, respectively.

MARS is the fourth planet from the sun and is a cold and dry planet. The dust which makes up the surface of the planet is a form of iron oxide, which is also what gives the planet its red color. The topography of Mars is very similar to that of Earth. Ice is located in some locations on Mars. Additionally, though the atmosphere is currently too thin for water to exist on the surface in liquid form, it is theorized that, in the past, water was abundant on the planet.

JUPITER is the largest planet in the solar system. It is the fifth planet from the sun. The planet itself is primarily gaseous, being composed of hydrogen and helium (along with others, in smaller amounts). The Great Red Spot is an enormous storm which has been ongoing on the planet for hundreds of years. The planet has dozens of moons and a very strong magnetic field.

The sixth planet from the sun, SATURN is primarily known because of the rings which orbit it. The planet is primarily gaseous, being comprised of helium and hydrogen (predominantly). The planet has multiple moons.

URANUS is the seventh planet from the sun. The equator of Uranus is at a right angle to the orbit of the planet, leading to it appearing to be on its side. This planet is about the same size as Neptune. Uranus has faint rings, multiple moons, methane, and a blueish-green tint.

Eighth planet from the sun, NEPTUNE is characterized by its cold temperature and its very strong winds. The core of Neptune is rocky. Neptune is about seventeen times the size of Earth. The planet was originally discovered using the power of math, by theorizing about the irregular orbit of Uranus being the result of a gravitational pull.

In recent years, PLUTO, formerly the ninth planet from the sun, has been reclassified from a true planet to a dwarf planet. The ASVAB is unlikely to ask any tricky questions about Pluto, especially considering it has only recently changed classifications. Pluto is smaller than the moon of the Earth, and it has an orbit near the outer edge of the local solar system. Its orbit around the sun takes around 248 years on Earth. The planet itself is rocky and very, very cold. Its atmosphere, what little there is, is extremely thin.

Oceanography

About 71 percent of the surface of the Earth itself is water. Relatively speaking, the water layer is not very thick in most places. The water contains minerals and salts which have been dissolved, and the vast majority of the water is held in four ocean basins. Seas, porous rocks, ice caps, lakes, and rivers contain the rest of the water on Earth. From smallest to largest, the oceans are the Arctic Ocean, the Indian Ocean, the Atlantic Ocean, and the Pacific Ocean.

The oceans play a very large role in the maintenance of the environment of the Earth. Things such as the shape of the Earth, the axial tilt, the way it rotates, and the relative distance from the sun affect the heating across the oceans. The water allows that heat to be distributed around the planet. That is also what helps to create ocean currents. Oceans are able to absorb and release heat and, thus, they help to regulate both the climate and the weather.

Ocean basin is a catch-all term for the land which is under the ocean. Generally speaking, these are comprised of basaltic rock. Some areas of the basin have a large amount of both seismic and crustal activity, however, and they are the areas which are responsible for the activity studied in plate tectonics.

Oceanography itself is a pretty broad topic and is a term used to describe the study of the ocean as well as oceanic ecosystems, currents, fluid dynamics, plate tectonics, and marine organisms. Typically, oceanography is going to fall into a multidisciplinary area. One of the major areas of study in oceanography, currently, is the study of the acidification of the ocean, which is a term used to describe the way that the pH of the ocean is decreasing due to carbon dioxide emissions into the atmosphere of the Earth.

Geology

The study of geology is the study of the Earth itself. The Earth, as you may know, is a planet which is revolving around the sun. The Earth is not uniform in shape (it is not a perfect sphere). It has a number of layers and a disparate topography. Among the layers of the Earth are the crust, the asthenosphere, the mantle, and the core. The thinnest of those layers is the crust, which is also the outermost layer (and the layer that humanity calls home). The crust of the Earth is about eight thousand miles around (in diameter). Together with the topmost layer of the mantle (the most solid part of the mantle), the two are known as the lithosphere.

The way that the interior of the Earth functions is a sub-discipline of geology known as seismology, which is the study of the seismic activity of the Earth. Seismic activity, for all intents and purposes, is another term for earthquakes.

The crust is not uniform at all and, in fact, has a unique topology. Think about the crust and what you know about it. Valleys, mountains, plains, oceans, etc. These are all variations in the height and thickness of the crust. The portion of the crust which forms continents is primarily granitic rock. On the other hand, the part of the crust which is located underneath the oceans is primarily comprised of basaltic rock. These two types of crust are then broken up into tectonic plates which move over the asthenosphere.

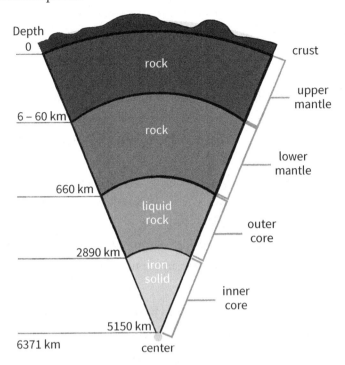

Figure 4.7. Layers of the Earth

The movement of the tectonic plates on each other is known as (big surprise) plate tectonics. This is what explains many natural phenomena that happen on the Earth, including volcanoes, how mountains are made (and destroyed), the way the sea floor changes, trenches in the ocean, and earthquakes. The crust is usually only stable for a small time frame, geologically speaking. Earthquakes always occur, volcanoes become active and go dormant, and many other things are ever-changing on the crust of the Earth.

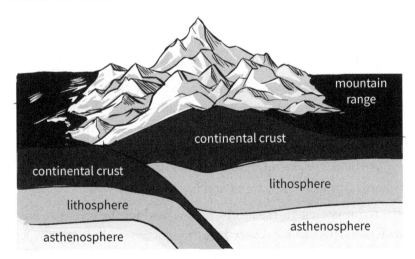

Figure 4.8. Earth's crust

Rocks on the Earth are also changing. Generally, rocks are made up of a combination of different inorganic crystalline substances known as minerals. Each type of mineral will have a specific chemical makeup and will have properties that are unique to them. Here are some of the most common minerals that you might encounter:

- **BAUXITE** is a type of rock composed primarily of aluminum oxides which have been hydrated.
- **QUARTZ** is the most abundant mineral in the crust of the Earth. This is the most common and simple form of all silicates. It is an oxide of silicon.
- **TALC** is a common and soft mineral which can be scratched with a fingernail.
- **PYRITE** is a mineral comprised of iron and sulfur. Pyrite is commonly known as "fool's gold" because of its resemblance to its namesake.
- **GRAPHITE** is a form of carbon. You may recognize this as being used in pencils and some other commercial applications.

Also included in this list would be all of the types of precious stones you can think of: emerald, ruby, opal, diamond, sapphire, etc.

Minerals comprise the types of rock that make up the Earth as well. The most common types of rocks are igneous, sedimentary, and metamorhic. **IGNEOUS ROCKS** have been formed through the melting and cooling of minerals within the mantle beneath the surface of the Earth. These can surface, initially, as lava. Some common examples of igneous rocks are granite, basalt, and solid volcanic lava.

SEDIMENTARY ROCKS have been created by the deposit of their composite materials onto the crust of the Earth and inside the bodies of water that help make up its surface.

METAMORPHIC ROCKS are rocks that already exist, but have gone through a transformation. When rock is put under immense heat and pressure, changes can occur within it. Some of the examples of this type of rock would be slate, gneiss, and marble. The changes occurring within the original rock can be both physical and chemical in nature.

There are two layers which are not layers of the planet itself, but which the crust is nevertheless in contact with—atmosphere and hydrosphere. **ATMOSPHERE** is a term for the layer of gasses which surround planets (or any large body with a significant amount of mass) which are being held by the gravity of those planets. The atmosphere of the Earth is primarily nitrogen and oxygen and has multiple layers.

The **HYDROSPHERE** is the term which is used for the collective water of the Earth. This includes everything from oceans to lakes, ponds, and rivers. On the Earth, the hydrosphere is about 70 percent of the surface.

The way that these two layers meet with the crust results in an energy exchange which manifests itself through three processes: weathering, erosion, and deposition.

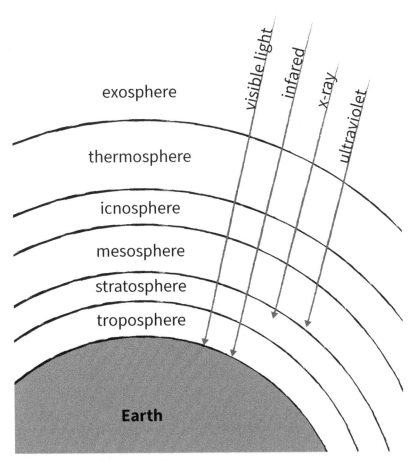

Figure 4.9. Earth's atmosphere

WEATHERING is the process of rocks and soil breaking down through contact with the atmosphere and waters of the Earth. This occurs without movement. It is important to note that this is not the same thing as erosion.

EROSION is the process of rocks and soil breaking down and being moved and deposited somewhere else through nature processes such as the flowing of water, wind, and storms.

DEPOSITION is a process through which soil and rocks are added to a mass through transport as a result of erosion and a loss of kinetic energy. Deposition would be rocks breaking off of a mountain because of a hard storm and then "depositing" down at the bottom of the mountain when they no longer have enough kinetic energy to move.

Through these processes, the crust of the Earth is always being changed. It is changing and being changed by natural forces.

Meteorology

METEOROLOGY is the study of the atmosphere, weather, and climate of the Earth. The atmosphere of the Earth is meant to help reflect, refract, and absorb the light energy being emitted from the sun. The process of these activities taking place is what leads to the energy flow that causes the weather and climate of the Earth.

WEATHER is a temporary, day to day, type of atmospheric condition. Think about rainstorms, sunny days, cold days, or snow. All of these are, generally, temporary conditions. They are not long-term, geologically speaking.

A CLIMATE is the longer term version of weather. Think about how California is generally hot and dry. How Florida is hot and wet. The northern states are cold. Jungles are hot and wet. Deserts are dry. All of these are types of climates. They are not temporary like normal weather would be.

The atmosphere of the Earth consist of many elements, but it is primarily composed of nitrogen and oxygen. About 1 percent of the atmosphere is made up of other gasses, such as carbon dioxide, ozone, and argon. As far as nitrogen is concerned, it makes up around 78 percent of the atmosphere. Oxygen makes up 21 percent.

There are multiple layers of the atmosphere, all of which are separated from each other by pauses (which have the largest variation in characteristics): exosphere, thermosphere, mesosphere, strasphere, and troposphere.

The EXOSPHERE is the farthest out part of the atmosphere. This is where satellites are orbiting the planet and where molecules have the potential to escape into space itself. The very bottom of the exosphere is known as the thermopause. The thermopause is about 375 miles above the surface of the Earth. The outermost boundary of the exosphere is about 6,200 miles above the surface of the Earth.

The THERMOSPHERE is the next layer when coming toward Earth from space. It is just inside of the exosphere, separated from it by the thermopause. This layer is between 53 and 375 miles above the surface of the Earth. This layer is known, colloquially, as the upper atmosphere. The gasses in this layer are very thin, but they become more and denser the closer you get to the Earth. This is the layer which absorbs the energy coming in from the sun (particularly ultraviolet radiation and x-ray radiation). That energy is what leads to the high temperature. The top of this layer is around –184° Fahrenheit while the bottom is around 3,600° Fahrenheit.

Between thirty-one and fifty-three miles above the surface of the Earth, lies a denser layer of atmosphere and gasses called the MESOSPHERE. The temperatures in this layer are around 5° Fahrenheit. Gasses here are usually thick enough to stop most small meteors that enter the atmosphere of the Earth, causing them to burn up. This layer, along with the stratosphere, are collectively known as the middle atmosphere. The boundary between the two layers is called the stratopause.

The STRATOSPHERE is from thirty-one miles above the surface of the Earth to around eight miles above the surface of the Earth, give or take four miles. About 19 percent of the gasses in the atmosphere are contained in the stratosphere, which has a very low water content. The temperature of this layer increases along with the height. The heat is a byproduct of the creation of ozone in this layer. The barrier between this layer and the troposphere is the tropopause.

The TROPOSPHERE is the lowest layer. This is where weather takes place. It goes from the surface of the Earth to between four and twelve miles above the surface. The exact height of this layer varies. The density of gas in the troposphere decreases with the height, and the air becomes thinner (which is why mountaintops have thin air).

The layer which humans live in is known as the troposphere, which is also the layer in which weather takes place. Some of the variables which are governed by weather conditions include the weight of the air (barometric pressure), air temperature, humidity, air velocity, clouds, and the levels of precipitation. Instruments are commonly used to help determine the relative levels of all of these. Below are some common weather measurement instruments:

- A THERMOMETER is a tool which is used to measure the temperature of the air using either mercury or alcohol.
- A BAROMETER would be used to measure the pressure of the air. If the pressure is going up, calm weather is coming. If the pressure if going down, expect rain.
- A PSYCHROMETERS is a device used to measure relative humidity through the use of evaporation.
- An ANEMOMETER is an instrument which measures wind speed using a series of cups which catch the wind and then turn a dial.
- A RAIN GAUGE is used to determine how much rain has fallen in a given period of time.
- WIND VANES are weather instruments which are used to help determine the direction of the wind.
- A HYGROMETER is an instrument which is used to help measure the humidity of the air.
- WEATHER MAPS show the atmospheric conditions over geographical areas. You are likely familiar with these from your local news.
- A COMPASS is an instrument which is used primarily for navigation and can be used to help determine directions.
- WEATHER BALLOONS are commonly used to help measure the conditions of weather in the upper atmosphere.

Obviously you will probably not have all of these in your home. It is likely, however, that you know of most of them (even if you do not know their names). It is also likely that you have used a few (or seen them used) at some point in your life.

Most meteorologists will utilize readings from a weather map and a number of these tools in order to determine what air masses are moving around in the troposphere. Air masses would be defined as sections of air that has a uniform content of moisture and temperature. The way that these air masses move and interact with each other determine the weather and, thus, are what most meteorologists use when determining changes in the weather.

The climate of the Earth and the seasonal changes which regularly occur are a result of the interaction between local water sources, the way that the Earth is tilted toward the sun, the altitude of the location in question, and the latitude of the location in question. These events are, obviously, very complex. It is not an easy task to help determine how the climate will change over time.

Tips

Below are some tips which will help you get through the general science subtest of the ASVAB:

- Don't spend too much time on any single question. If you don't know it, make an educated guess and move on.

- Try to eliminate ridiculous answers immediately. Usually, two or possibly three of the answers for any given question will be so wrong you don't even need to evaluate them for any significant amount of time.

- It will not do you any good to try to learn everything there is to know about science right off the bat. Instead, study in bite sized chunks and review regularly. If you find yourself weak in one or more areas, then spend a little extra time on them.

- Creating acronyms and tongue twisters can immensely help when attempting to remember all of the facts and figures in this section.

Practice Questions

1. The exosphere is the outermost layer of the atmosphere. How far above the surface of the Earth is the outermost boundary of the exosphere?

    A. 6,200 miles

    B. 200 kilometers

    C. 100,000 miles

    D. 5,000 miles

2. Sea water is composed primarily of water (H_2O) and _____.

    A. fish

    B. potassium chloride

    C. gold

    D. sodium chloride

3. This organelle is known as the powerhouse of the cell. It is where ATP is first created inside cells.

    A. plasma membrane

    B. mitochondria

    C. nucleus

    D. ribosome

4. In our local solar system, there are eight planets. What do these plants rotate around? What is the body at the center of the local solar system?

    A. sun

    B. moon

    C. Earth

    D. Venus

5. What is the term which describes interactions between communities of organisms and the environment that they live in?

    A. foot system

    B. community

    C. ecosystem

    D. environment

6. Toothpaste is a compound which is used to help prevent the decaying of teeth. What element is added to toothpaste in order to accomplish this?

    A. fluoride

    B. calcium

    C. silver

    D. amalgam

7. The Earth is made up of multiple layers. What is the outermost layer of the Earth itself?

    A. core

    B. crust

    C. mantle

    D. oceans

8. Which kingdom is the one in which ameba belong?

    A. homo sapiens

    B. fungi

    C. protists

    D. viruses

9. Many chemicals are solvents. One chemical is known as the *universal solvent*. Which chemical is it?

    A. water

    B. ammonia

    C. bleach

    D. sodium chloride

10. All cells have some sort of genetic material in them. In eukaryotic cells, where would the genetic material (DNA) be found?

    A. ribosomes

    B. mitochondria

    C. liposome

    D. nucleus

11. There is more than one type of nervous system in the human body. Which one is the brain a part of?

    A. peripheral nervous system

    B. central nervous system

    C. nerve core

    D. fascial nervous system

12. Plant cells and animal cells share some characteristics and differ in others. Which is the best example of a process in plants but not animals?

    A. making energy

    B. ATP synthesis

    C. photosynthesis

    D. replication

13. What is the branch of Earth science concerned with the atmosphere and weather?

    A. geometry

    B. atmospherology

    C. meteorology

    D. astrology

14. What are amino acids used to create?

    A. cytoplasm

    B. DNA

    C. cells

    D. proteins

15. Out of the planets, one has a "vertical" equator when compared to the others. Which one is it?

    A. Uranus

    B. Neptune

    C. Venus

    D. Jupiter

General Science Answer Key

1.	A.	9.	A.
2.	D.	10.	D.
3.	B.	11.	B.
4.	A.	12.	C.
5.	C.	13.	C.
6.	A.	14.	D.
7.	B.	15.	A.
8.	C.		

GO ON

Review

Basic Principles

This includes the metric system, conversion back and forth, basic temperature information, and other commonly used scientific jargon which is included in all of the other sections of the subtest (and a number of others).

Scientific Method

The scientific method is the means by which nearly all scientific discoveries are made. The steps are simple: observe, hypothesize, predict, experiment, and repeat.

Disciplines

The disciplines, broadly, which are included on the general sciences subtest are life science, chemistry, physics, and earth science.

Life Science

Biology and the fields of science which pay close attention to the study of living organisms.

- CLASSIFICATION is the way that organisms are grouped together for description and cataloging.
- The THEORY OF EVOLUTION describes the way that organisms and systems can change over time as a result of the environment that they are found in.
- CELLS are the basic unit of life. This section talks about cells, organelles, and the way that cells interact with each other.
- RESPIRATION, FERMENTATION, PHOTOSYNTHESIS are the three methods through which cells are able to create energy.
- CELL DIVISION is the study of the replication of cells, both mitosis and meiosis are included.
- GENETICS is the study of genes and how they function. DNA to RNA to protein is the central dogma of modern genetics.
- ANATOMY is the study of the human body and how it functions, including the many parts of the body.
- ECOLOGY is the study of environments and the interaction of plants and animals within those environments.

Chemistry

The type of physical science which is primarily concerned with the structure, properties, and composition of matter itself.

- ATOMIC STRUCTURE means atoms, their composition, and the subatomic particles that make them up.

- **ATOMS** are the most basic form of elements; **MOLECULES** are combinations of one or more types of atoms, and **COMPOUNDS** are combinations of molecules.
- The **PERIODIC TABLE** is the best way to list all of the elements that exist along with basic information about each of them.
- **REACTIONS AND EQUATIONS** explain the way chemical reactions are written, balanced, and explained.
- **SOLUTIONS, ACIDS, AND BASES** are mixtures of chemicals; information about how hydrogen or hydroxide interacts inside mixtures.
- Information about the most important **ELEMENTS** that are found in nature.
- **ORGANIC CHEMISTRY** is the study of compounds and molecules which contain carbon.
- Information about **METALS** and the characteristics of common metals that are found in nature.
- **ENERGY AND RADIOACTIVITY** as it relates to chemicals and bonds between molecules and atoms.

Physics

A natural science which is concerned with the study of matter and how it moves through time and space.

- **MOTION** is the study of changes in position or time and the study of movement itself.
- **KINEMATICS** is a subdiscipline of mechanics which is concerned with the motion of objects without considering the cause of the motion.
- The **LAWS OF MOTION** as outlined by Isaac Newton. These define the way that objects and forces acting on them relate to each other.
- **KINETIC AND POTENTIAL ENERGY** as it relates to physics.
- **FLUID MECHANICS** is how fluids move and react to their environment.
- **ELECTRICITY** is a basic primer on the physical basis of electricity, including the interactions between subatomic particles which lead to its manifestation in the environment.
- What **SOUND** is, the way sound travels, and how sounds are amplified when they need to be.

Earth Science and Astronomy

Primarily the study of the Earth itself and the way that it relates to other bodies in the universe.

- **ASTRONOMY** is the study of the stars, the planets, and space.
- **OCEANOGRAPHY** is a multidisciplinary study of the oceans. This includes the currents, marine biology, plate tectonics, and other subjects involving the oceans.
- **GEOLOGY** is the study of the Earth itself, including the types of rocks it is composed of and the way that the rocks change over time.
- **METEOROLOGY** is a term used to describe the study of the local and macro-climates of the Earth, including the weather.

five

ARITHMETIC REASONING

Introduction

The **ARITHMETIC REASONING** subtest of the ASVAB is primarily comprised of mathematical word problems. The general purpose of this test is to determine how well you can apply your mathematical knowledge to situations you may encounter in the real world. It is important to note that the questions you encounter on this section of the test involve both reasoning (logic) and arithmetic. They are not as complicated as the ones that you might encounter in the other math portion of the ASVAB, but they may be complicated since the actual problem may not be directly stated.

This is one of the only test sections you are likely to encounter that will necessitate the need of the scratch paper that you have been provided. You will NOT be allowed to use a calculator, so you should practice these types of problems without the use of one to be prepared.

Qualification for many jobs within the military depends on your success on the arithmetic reasoning test as well. The following line scores utilize the arithmetic reasoning score:

Table 5.1. Line scores by military branch

BRANCH	LINE SCORE
Army	Clerical, general technical, skilled technical, operators and food, surveillance and communications
Marines	General technical
Navy/Coast Guard	Administrative, health, nuclear, general technical
Air Force	Administrative and general

There are three types of questions you are likely to encounter on the arithmetic reasoning section of the ASVAB: algebra word problems, fact-finding word problems, and geometry word problems.

ALGEBRA WORD PROBLEMS are word problems that necessitate the creation of an equation with an unknown in order to solve them. These are not too complex and, generally speaking, you will not have to use too much algebra to solve these.

EXAMPLE

John spent $42 on a pair of shoes. That was $14 less than twice the amount of money John spent on a new pair of Jeans. How much did the jeans cost?

As you can see, you are trying to find a value that isn't here. You would best set this problem up in the following way, the unknown being x:

$2x - 14 = 42$

In other words, 2 times x (the value of the jeans) minus 14 will yield **42**, the value of the shoes that John bought. Do you see how you can "translate" the word problem into math, almost word by word? And at this point, you'll probably want to plug in the answer choices to see which one works. **The correct answer is 28**, since 2 × 28 − 14 = 42. You can also solve this using more complicated algebra, as we'll see in the mathematics knowledge section.

The second type, **FACT-FINDING WORD PROBLEMS**, are problems that state facts and then ask you to do something with them.

EXAMPLE

The chess club is trying to raise money by selling chocolates that are shaped like chess pieces. Jimmy sold 3 of them, Sally sold 5, and John sold 12. How many chocolates were sold in total?

First, determine the facts:

Jimmy = 3 chocolates

Sally = 5 chocolates

John = 12 chocolates

Total chocolates sold = ?

Next, simply add them together.

3 + 5 + 12 = 22

20 chocolates were sold in total.

The third type, **GEOMETRY WORD PROBLEMS**, are problems asking you to find a perimeter or area of a given space. Typically, these will not be more complex than that.

With a firm grasp of these types of questions, you will be prepared for the arithmetic reasoning subtest questions.

Numbers

Whole Numbers

Most people know what whole numbers are, even if you do not know the term "whole number". A whole number is exactly what it sounds like: an *entire* number. Examples of whole numbers are 0, 1, 2, 3, 4, and so on. There is no decimal or fractional part. Whole numbers are part of a system that is known as the place value system. In this system, numbers are labeled with specific words, depending on their position.

You may have heard the term "counting numbers" or "natural numbers," which are whole numbers without the "0". Natural numbers are often used for basic counting, ordering, and things of that nature. These numbers are the simplest form of numbers, from which other number sets are created.

Here is a brief table you can use to refresh your memory and better grasp this concept:

Table 5.2. Whole numbers

VALUE	PLACE
one	1
ten	10
hundred	100
thousand	1,000
ten thousand	10,000
hundred thousand	100,000
million	1,000,000

This decimal number system (also called base 10 since it has 10 as its base) continues into the millions and beyond.

If you were given a number such as 14,567, the word associated with it would be fourteen thousand, five hundred and sixty-seven. It is important to grasp this concept because on the arithmetic reasoning portion of the ASVAB, you may encounter numbers as their long-form versions rather than the punctual versions you are used to.

ROUNDING is a term that is used to quickly determine the relative size of a whole number based on a new number (either larger or smaller than the original number). Rounding is the best method to approximate numbers quickly. Below is the general procedure for rounding:

- Choose the digit you will be rounding and underline it (so you know which one it is).
- Look at the digit directly to the right of the digit you are rounding. If that number is less than 5, then you will leave the underlined digit alone. If the digit to the right of the digit you are rounding is 5 or more, then you will add 1 to the digit you are rounding.
- Last, replace all of the digits that are to the right of the underlined digit with 0s (up to the decimal point).

EXAMPLE

Round 356,782 to the nearest ten thousandth.

First you will select the ten-thousands spot, the 5. Then you will look at the number to the right of it, the 6. That number is larger than 4, so you add one to the 5 spot, giving you 366,472. Now you will replace numbers to the right of the primary digit with 0's, giving you **360,000**.

Fractions and Decimals

Fractions and decimals are both methods of elaborating on *parts* of whole numbers. This is usually illustrated by two numbers, say 0 and 1, and then viewing them via a line between them. That line represents all of the fractions or decimals in between those two numbers.

Take FRACTIONS first. The most basic way to view a faction is to look at it as parts of a whole. The division sign ("/" or "−"), which separate the two numbers, can represent the words *out of.* So if you have $\frac{2}{4}$, you have 2 out of 4 parts, which is actually the same as 1 out of 2 parts. For example:

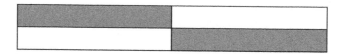

Figure 5.1. Two out of four parts

In the above figure, you can see that there are 4 parts (rectangular boxes). Of these, 2 of them are gray. So you have $\frac{2}{4}$ boxes that are gray.

Note that this also represents $\frac{1}{2}$ of the boxes, and we can say that $\frac{2}{4}$ reduces or simplifies to $\frac{1}{2}$.

The top part of the fraction is called the **NUMERATOR** and the bottom part is called the **DENOMINATOR**. When looking at fractions, if the numerator is less than the denominator, then it is referred to as a **PROPER FRACTION**. The value of these types of fractions is going to be less than 1. Some examples of proper fractions are $\frac{3}{9}$ or $\frac{1}{3}, \frac{2}{5}$, and $\frac{8}{19}$.

If the numerator is equal to or greater than the denominator, then you have an **IMPROPER FRACTION**. Some examples of improper fractions are $\frac{6}{3}$ or $\frac{2}{1}$ (which is 2), or $\frac{8}{3}$ and $\frac{5}{2}$. Any improper fraction can be written in the form of a **MIXED NUMBER**, and any mixed number can be written in the form of an improper fraction. For example, $\frac{4}{3}$ is the same as $\frac{3}{3} + \frac{1}{3} = 1\frac{1}{3}$.

It should be noted that with fractions, having $\frac{4}{4}$ is the same as having 1, $\frac{8}{4}$ is the same as 2, and $\frac{9}{3}$ is the same as 3; these pairs of fractions are **EQUIVALENT**. (See how if we were to multiply both the numerator and denominator by the same number, we will get equivalent fractions?) Likewise, having $\frac{6}{4}$ is the same as $\frac{3}{2}$ which equals $1\frac{1}{2}$. This is how fractions are converted back and forth into whole numbers and mixed fractions.

To review:

- Proper fractions: These are fractions with a value of less than 1. The numerator is smaller than the denominator. $\frac{1}{2}$ is an example.

- Improper fractions: These are fractions with a value of 1 or more. The numerator is larger than the denominator. $\frac{7}{5}$ is an example.

- Mixed numbers: These are whole numbers that are combined together with fractions. $5\frac{1}{2}$ is an example.

Now onto **DECIMALS**. Everything talked about earlier, in terms of whole numbers, have included digits that are powers of 10, but greater than 1 (to the left-hand side of the decimal point). Again, the number system is a **BASE 10 SYSTEM**. Decimals are a kind of shorthand that are used to describe base 10 numbers (fractions of a whole number) that are on the right-hand side of the decimal point, which means they are less than 1.

Here is a table which might help explain this concept a bit better. Notice that the names of the places to the right of the decimal include a *th* at the end:

Table 5.3. Decimals

PLACE	DECIMAL
tenth	0.1
hundredths	0.01
thousandth	0.001

In this way, you can describe numbers smaller than whole numbers in a method other than using fractions. This is also how percents are written (see the next section). So if you have 0.5, you have exactly $\frac{1}{2}$ of 1. You can also think of this as $\frac{5}{10}$, or $\frac{1}{2}$, or $\frac{.5}{1}$, which is five-tenths of 1.

Remember that typically in fractions and percent problems, *of* represents multiplication; for example, $\frac{1}{2}$ of 4 = $\frac{1}{2}$ × 4 = 2.

Percents

PERCENTS can be considered fractions that have a bottom number (denominator) equal to 100. You can think of this as *per hundred*, which is what *percent* actually means. The symbol that is used to denote a percent is %. For example:

$\frac{50}{100}$ is the same thing as .50, which is the same thing as 50%.

You can see that there is a very clear and definite relationship among fractions, decimals, and percents. Percents are typically used in order to describe portions of a whole, just like the other two are. If you have a fraction, you can convert the top and bottom numbers to $\frac{\#}{100}$ in order to determine the percent. Likewise, you can simply shift the decimal two places to the right to get a percent from a decimal; for example 0.75 is 75%. To get a decimal from a percent, shift the decimal two places to the left; for example, 45% is .45.

A typical question about percents on the ASVAB might be converting them to decimals and back to percents.

Operations

Operations are the calculations that are done to numbers. *The Fundamental Operations of Arithmetic*, as they are called, are addition, subtraction, multiplication, and division. Solving the problems in the ASVAB will require you to perform these operations on whole numbers, decimals, and fractions.

Addition and Subtraction

The ADDITION of whole numbers is not too complicated. Adding two numbers together is an operation that results in a total that is known as the sum. The very first step here is to line up the numbers you are going to add by their place value. All of the ones will go in a line (column), and then tens, hundreds, thousands, and so on. Once this has been done, you will add the columns together. Any time the sum of a column reaches 10, you need to carry over the 10 as a single digit to the next column to the left. So 10 in the ones column would be 1 in the tens column. Proceed to do this until the numbers have been added up.

EXAMPLE

Add together 20, 15, 2, and 107.

20

15

2

+107

144

Now look at those numbers. The digits in the ones column (the right-most column) add up to 14. Bring down the 4 to the ones digit and carry over the 10 to the next column (to the left). So 2 + 1 + 1 (the carryover) + 0 = 4. So there is a 4 in the tens digit as well. The hundreds column only has a 1. So the final result is **144**.

Adding together decimals is the exact same procedure. The only change is that you have to deal with that pesky decimal point. All you have to do is line up the decimal point vertically with every single number you need to add. You then add them together just like before to get your sum. For whole numbers, you can make this easier by adding a decimal point to the right of the number (so 17 would become 17.0 or 17.00, for example, depending on how long the other decimals are).

If you want to make things easier, you can just put 0s in the place of any empty spots to make it easier to add together.

EXAMPLE

Add together 13.2, 0.54, 2, and 0.008.

13.2

0.54

2

+ 0.008

The best way to do this is to modify the numbers a little bit. Note that they are still the same numbers, but easier to add:

13.200

00.540

02.000

+ 00.008

15.748

You can readily see that this small procedure makes these numbers much easier to add together. Adding them will give you **15.748.**

SUBTRACTION is the inverse or opposite of addition. It could also be considered another form of addition (by adding negative numbers),

65

.25

125

1560

16.25

but that concept will be covered in a later section. Subtracting a number from another number will give you a result that is known as the difference. The steps are the same but backward. Line up the two numbers like before. (You should only subtract two numbers at a time). If the number on top is less than the one on the bottom for that column, then "borrow" from the next column (the one to the left) and continue with the operation.

EXAMPLE

Subtract 108 from 110.

```
  110
 −108
    2
```

Start with the right-most column, the ones column. Since 8 is greater than 0, you have to borrow from the tens column to turn the 0 into a 10 before you subtract. But then you have to turn the 1 in the tens column into a 0 (subtract 1 in the tens column). So the final answer is just **2**. Again, like addition, this is a pretty simple process that most individuals are already used to doing without even thinking about it.

The subtraction of decimals, again, is much the same as the addition of decimals. Before subtracting, you have to line up the decimal points of the numbers that are being subtracted from one another and then place the decimal point below those two amounts in the same spot.

EXAMPLE

Subtract 89.3 from 109.4.

```
  109.4
 − 89.3
      2
```

The first thing you want to do is fill in any zeroes to the right of the decimal point. In this case, we don't need to borrow.

```
  109.4
 − 089.3
  020.1
```

Practice makes perfect when it comes to addition and subtraction. Spend a little bit of time working on these types of problems on your own. The more you do it, the easier it will become. It would benefit you greatly to do these types of problems on paper without the use of a calculator. This will help you with mental calculations and will greatly increase your accuracy. There are a lot of websites that provide practice on these types of problems.

Multiplication

MULTIPLICATION is the next basic operation that you will need to know for the ASVAB. The multiplication of two numbers, which is basically repeated addition, will result in a number that is known as the product. The very first step in multiplication is putting the number with the most digits on top and lining up the numbers to the right. There are two general situations you will encounter when you have to multiply two numbers together.

1. One of the numbers you have to multiple has only a single digit. For example multiplying 15 by 8.
2. Both of the numbers you have to multiply have more than one digit. For example multiplying 10 by 16.

In the first situation, you want to put the number with the single digit on the bottom and then multiply all of the digits of the top number by the bottom number. You will begin on the right-hand side of the top number and then multiply individual digits, working to the left. If the multiplication results in a number with more than one digit, you will write down the ones digit in the product and carry the tens digit to the next column. Then you will add the carried number to the next multiplication.

EXAMPLE

Multiply 123 by 6.

123

× 6

Notice how the number with the most digit is put on top. The first step would be to look at the singles digit (6). Multiple that number (6) by the rightmost number in the top column (3). That will result in 18, so we'll put an 8 below. Moving left, the 6 is multiplied by 2 (on a diagonal), resulting in 12, but we also have to add the 1 that we just carried, so we get 13. Put a 3 below, and carry 1. Now the 6 is multiplied by the 1 to get 6, but we need to carry over the 1 from the 13, so we get 7. The final result is 738.

123

× 6

738

The second situation is the same as the first, but with an additional step. Again, remember to place the number with the most digits on top, and start multiplying from the right (the ones digits). You will first multiple the ones column by the top number, then the tens, then the hundreds, and so on until you are finished. Each value place you move to the left will add a zero to the right of the product. The products will then be added together to give the final result.

Multiply 201 by 13.

$$
\begin{array}{r}
201 \\
\times\ 13 \\
\hline
603 \\
+\ 2010 \\
\hline
\mathbf{2613}
\end{array}
$$

The first thing that has so be done after arranging the numbers in the correct way is to multiply the 3 by 1. That will give you 3. Then 3 by 0 and 3 by 2 respectively, for a result of 603 total. Now you will do the same thing with the 1 (the tens digit of the second number). 1 by 201 is, obviously, 201. But since you are using the tens digit, add a 0 to the right-hand side of this number, giving you 2010. This is to reflect the higher value of the digit being multiplied. Now, finally, you will add the 603 and 2010, giving you a final result of **2613.**

The multiplication of decimals requires something quite different than what you did during the addition and subtraction. In this situation, you will completely ignore the decimal point altogether. Multiply the two numbers together as if they were whole numbers. Once you are done doing that, you will figure out where to place the decimal point back in. This sounds much more difficult than it actually is.

Multiply 54.8 by 9.9.

$$
\begin{array}{r}
54.8 \\
\times\ 9.9
\end{array}
$$

First, you will take out the decimal points so that the multiplication will be a bit simpler.

$$
\begin{array}{r}
548 \\
\times\ 99
\end{array}
$$

Now finish up the multiplication and addition:

$$
\begin{array}{r}
548 \\
\times\ \ 99 \\
\hline
4932 \\
+\ 49320 \\
\hline
54252
\end{array}
$$

To deal with the decimal points, add up the number of digits to the right of the decimal points in the original numbers. This does not mean to add up the actual numbers; it means add the number of total digits to the right of the decimal points. So 54.8 has one digit to the right of the decimal point and 9.9 has one digit to the right of the decimal point. This means you have 1 + 1 digits to the right of the decimal points (2). In the final product, 54252, just stick the decimal point in (from the right) based on that number. The result would be **542.52**.

This process is a bit more complicated than the simple addition and subtraction that was done earlier and it is also more complicated than the multiplication of whole numbers. Regardless, it will get a lot easier once you have done it a few times.

Division

DIVISION is the process of splitting a number into parts or groups, and this yields a quotient. There are four basic steps that you will undertake to do long division and you will repeat these steps for every division operation that you have to perform.

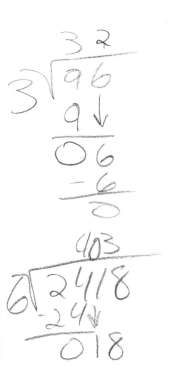

EXAMPLE

Divide 5 into 65.

The first step here is going to be to set up the problem:

$\frac{65}{5}$ or $5\sqrt{65}$ fd

Now you will begin the actual division. First, select the leftmost digit of the number being divided (the dividend) and see how many times the divisor goes into it. In this case, 5 goes into 6 one (1) time. 6 – 5 is 1, leaving a remainder of 1. Now you will take the remainder, along with the next digit in the number being divided, in this case 1 and 5 respectively, and see how many times the divisor goes into it. How many times does 5 go into 15? The answer is **3**. Pair that with the original digit you got and you have your solution: 13.

```
    13
5√65
    5 0
    1 5
    1 5
      0
```

To test this and see if you got the right answer your answer, multiply 13 by 5 and see what you get. Is it the original number, 65? Yes, it is. So the division is correct. Always test once this has been done, and remember to add any remainder you might get to the product of the divisor and dividend when checking your answer.

When doing this with decimals, you will work exactly as you did with multiplication. Ignore the decimals until the end and then deal with them just like before. Add the decimal in exactly where it would be in the number that is being divided (move the decimal straight up into the answer). However, if there is a decimal in the outside (divisor), you must move the decimal to the right to make this a whole number, and then do the same (same number of places) on the inside (the dividend). Then you move the decimal straight up.

Word Problems

Basic Problems

The most basic problems that you will encounter on the arithmetic reasoning section of the ASVAB are problems that require one or two steps to complete. Generally, the most difficult part about these types of problems is figuring out what they are actually asking you to do.

One or two step problems usually have only one computation that needs to be made. At times, you may have to do problems that are a bit more complex. With that being said, the question you are being asked will usually be clear and not hard to figure out. The way these problems will be discussed and taught here will involve logic and examples. The exact computations will be up to you to complete on your own. And take advantage of the fact that the answers are provided by plugging them in backwards, if you need to.

EXAMPLE

1. Sally is going to the store and she is buying her groceries for the week. She is buying some apples for $5.00, some pears for $4.25, and some grapes for $9.50. How much does she spend at the store during this trip?

   This is a very simple problem. It is one basic step and only involves basic arithmetic. Three numbers are provided: $5.00, $4.25, and $9.50. The question asks you how much she spends. The way to solve this is to sum the three numbers, giving you a total of **$18.75**.

2. John has a job laying tile with his father's company. He is paid $9 an hour. He works for two weeks per pay period. This period, he works 10 hours during the first week and 5 hours during the second week. How much did John make during this pay period?

   This problem is a bit more complex since two operations are used. The information you are given lays it all out: He gets $9 an hour, a pay period is 2 weeks, he worked 10 hours the first week and 5 hours the second week. That's 15 total hours. 15 hours times $9 per hour yields $135 for this two-week pay period. You can also solve this by multiplying 10 hours by $9 and then 5 hours by $9 and adding them together. The result will be the same, even though the process is different.

Percent and Interest

The most common types of problems that you will encounter in the real world (in terms of word problems for mathematics) are going to be dealing with money. This is when PERCENT AND INTEREST operations come into play. Either you will be dealing with interest that is being accrued on some amount of money in the bank or on a loan you are paying, or you will be trying to find a certain percentage of a sum. These two things are typically related.

Here are two things to keep in mind when you begin working on this type of problem:

- The first thing you need to do is to convert the percentage into a whole number or decimal. You can do this by removing the percentage and moving the decimal two places to the left. (Remember, when you are removing the percent, move the decimal away from it.)
- The second thing you need to do is multiply the decimal (originally the percentage) by the figure being calculated.
- Finally, if necessary, add the percentage amount (which is now a whole number or decimal) to the original amount.

EXAMPLE

You have an investment account and you have put in $6,000. The account earns 7.5% interest annually. How much money do you have, in total, when one year is up?

The first thing you want to do is determine the important numbers here. Beginning amount = 6000, 1 year = 7.5% interest. Interest interest for one year = 7.5% (or 0.075).

You can do this in one of two ways. You can set it up like a standard multiplication problem and then do the addition, or you can set it up algebraicallyin one step. Here is each method in fullthe first way:

$6000 × 0.075 = 450

$6000 + $450 = $6450

In this method, you do the multiplication to get the interest added for one year, and then add the total of the interest for one yearthis to the original sum.

The second method involves a bit of algebrasaves a step by combining the interest and the original amount:. In this one, t = time (in years).

$6000 × 1.075 = **$6450**

Again, when you are dealing with percentages, it is important to convert them in the proper way by moving the decimal two places to the left when removing the percentage; for example, 10% is 0.10 and

5% is 0.05. Percentages between 1% and 99% take up the first two digits to the right of the decimal point in a conversion.

Ratios and Proportions

RATIOS involve comparisons between numbers, and PROPORTIONS involve the equalities of ratios. Ratios are usually written with a ":" separating them. So a 5 to 1 ratio would be written 5:1. These can be written fractionally as well, which is useful when you begin working with some of the word problems that involve actual proportions. For example, 5:1 would be $\frac{5}{1}$. Comparing two numbers using a ratio results (with a 1 on the bottom typically) is what is known as a unit rate. This is where you get terms like "miles per hour" or "kilometers per hour" since "per" can mean "divided by" (this is the same as finishing the division of a fractional form of a ratio).

EXAMPLE

You want to figure out how much you are paying for your flour. On the price tag, it says that the flour is $9.95 and that the bag contains 6 ounces of flour. How much does the flour cost on a per ounce basis?

You want the cost of one ounce of flour, but you are given the cost of six ounces. Turn this rate into a fraction, and then set up a proportion with ounces on the left and money on the right: $\frac{1}{6} = \frac{x}{9.95}$. You can cross-multiply to get $6x = 9.95$, so the final result is **$1.66** (rounded to the nearest cent).

All problems of this type will be solved in roughly the same way, as you will see when you begin doing them for yourself.

Problems of Measurement

Measurement problems are extremely simple. You are going to, almost universally, be asked to determine either the area or the perimeter of a given shape. The perimeter is the distance around the geometric shape. The area is the total amount of space inside of the geometric shape.

This is the formula for finding perimeter, where l = length, w = width, and P = perimeter:

$$P = l + l + w + w$$

That can also be written as $P = 2l + 2w$. It is both of the lengths added with both of the widths. The formula for area, where l = length, w = width, A = area is:

$$A = l \times w$$

Area is length times width.

Tips

Here are some tips to help you with the arithmetic reasoning subtest of the ASVAB:

- ◆ Make sure you read the entire question before you attempt to answer. It is extremely important to confirm what the question is actually asking you, since with multiple choice problems, they will try to trick you.

- ◆ The answer should, at a glance, make sense based on what you are asked for. If you are asked how many eggs someone is carrying home from the store, the answer is going to likely be in the range of 0 – 100. So you can quickly eliminate any ridiculous looking answers early on.

- ◆ When you set up the equation that you will be using, be careful to put the correct numbers in the correct places. For example, when setting up proportions, make sure matching units are in the same spots (for example, on the left or on top).

- ◆ Generally, they will not give you figures if you don't need to use them. So if you see measurements and numbers in the problem, be sure to take note and figure out where you need to use them.

- ◆ Follow the order of operations. This shouldn't really even have to be said because it is so obvious, but always follow it. Some of the answers will likely be designed to trick you as if you hadn't done the math in the correct order.

- ◆ When it comes to basic arithmetic operations, you will find that practice makes perfect. The more you can practice, the easier the math problems will be. Make sure units match; for example, you may gave to convert some of the numbers from hours to minutes to match other numbers.

- ◆ If you can't figure out answer from the question, try each answer in the multiple choice to see if they work. In other words, you may have to work backwards for some problems.

- ◆ Try simpler numbers if you can't figure out the problems with the numbers given. For example, use $100 for percent problems to figure out the correct operation(s). Then apply these operations, using the more complex numbers in the problem.

- ◆ For word problems with variables representing real numbers, use real numbers to get an answer (to see what you're doing) and then put variables back in.

- ◆ Draw pictures!

Practice Questions

2y+3x = 55
y + 10 = x

1. Which of the following two integers have a sum equaling a number less than 149?

   A. 89, 65 = *154*

   B. 55, 93 = *148*

   C. 93, 57 = *150*

   D. 84, 77 = *161*

2. John is going to a big party tonight and he needs to buy a new dress shirt for the occasion. The last time he went to the store, he bought one for around $20. The price is now $28. What percentage of the original price has it increased?

   A. 30%

   B. 40%

   C. 45%

   D. 50%

   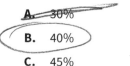

3. Rebecca, Emily, and Kate all live on the same straight road. Rebecca lives 1.4 miles from Kate and 0.8 miles from Emily. What is the minimum distance Emily could live from Kate?

   A. 2.2 miles

   B. 1.1 miles

   C. 0.8 miles

   D. 0.6 miles

4. One of the classes in a local elementary school has *g* girls. That is three more than four times how many boys are in that class. How many boys are in the local elementary school class? You may use *x* to represent the number of boys.

   A. $x = \frac{g-3}{4}$

   B. 13

   C. 12

   C. $b = \frac{4}{g-3}$

   g = 4b + 3

   $\frac{g-3}{4} = b$

5. One number, *y*, is 10 more than a second number. If you double that second number and add triple the higher number, the resulting number is 55. What are those two numbers?

   A. 2, 12

   B. 10, 12

   C. 5, 15

   D. 5, 10

   y + 10 = 2y + 3x

   2y + 3(y+10) = 55
   5y = 25   y = 5

6. Convert 28% to a decimal.

   A. 2.8

   B. 0.028

   C. 0.0028

   D. 0.28

7. A man has a set of jacks. He has *j* jacks, which is twice as many jacks as he has balls (*b*). How many balls does he have?

   A. *b = 2j*

   B. $b = \frac{j-4}{2}$

   C. $j = \frac{b-4}{2}$

   D. $b = \frac{j}{2}$

   2j = b

8. A class has around 30 students in it. Those students are divided into three groups. The first group has the most students in it. The second and third group both have 8 students each. How many students are in the first group?

   A. 14

   B. 15

   C. 8

   D. 16

9. John is a tractor driver. He works for $20 an hour and he gets paid an additional $5 for every acre of land he works. John is hired to work on a small farm. The work takes him 6 hours and he manages to cover all 10 acres of the farm. How much money is John going to be paid for this job?

   A. $120
   B. $50
   C. $150
   D. $170

10. You are doing some yardwork and you need to put seed onto your lawn. One bag of seed will cover 15 square feet. How many bags are you going to need in order to cover a yard that is 5 feet by 6 feet?

    A. 2 bags
    B. 3 bags
    C. 5 bags
    D. 15 bags

11. You are buying supplies for your class cookout. You are expecting 32 students and 2 teachers to attend. You estimate that for every 3 people, you will need 5 hamburgers. The hamburger patties come in packs of 6. How many packs should you buy?

    A. 10 packs
    B. 9 packs
    C. 8 packs
    D. 4 packs

12. Mark has a yard that is 18 feet long by 30 feet wide. What is the perimeter of the yard that Mark has?

    A. 48 feet
    B. 96 feet
    C. 540 feet
    D. 12 feet

13. A large construction company is trying to lay concrete for a new apartment building. They charge by the square foot. The price per square foot is 35 cents. The apartment building is going to be 500 feet long by 325 feet wide. How much is the total cost going to be for laying the concrete?

    A. $56,875
    B. $170,625
    C. $16,350
    D. $162,500

14. Sally is working a new job. She gets paid $10 per hour for stocking the shelves at a local grocery store. She is paid weekly and this is the week that her bills are due for the month. Her rent is $100, her car payment is $50, and her groceries cost her $35. She worked 40 hours this week. How much money does she have left over in her paycheck after she pays all of her bills?

    A. $394
    B. $185
    C. $400
    D. $215

15. You are looking to buy a home. The price of the home is $135,000. Every year, you will have to pay 2.5% of the value of the home in taxes. After 3 years, how much money will you have paid in taxes on the home?

    A. $3,375
    B. $13,375
    C. $10,125
    D. $6750

Arithmetic Reasoning
Answer Key

1.	B.	9.	D.
2.	B.	10.	A.
3.	D.	11.	A.
4.	A.	12.	B.
5.	C.	13.	A.
6.	D.	14.	D.
7.	D.	15.	C.
8.	A.		

Review and Takeaways

In the arithmetic reasoning portion of the ASVAB math section, you are expected to know basic algebra and arithmetic operations. You are being tested on your ability to use logic to solve these word problems and to pick out the important information. The key here is to separate the "signal from the noise" in the answers.

Review

NUMBERS: This section covered the basic number system, including whole numbers, partial numbers, and percentages.

- **WHOLE NUMBERS:** Whole numbers are the numbers commonly used to count, and include the number "0". They contain no fractional or decimal portions.

- **FRACTIONS AND DECIMALS:** Fractions and decimals are numbers that exist between two whole numbers.

- **PERCENTS:** Percentages are another way of saying "per 100". These are another method of explaining partial portions of some whole, and they can be converted back and forth to fractions and decimals (and sometimes whole numbers).

OPERATIONS: The operations section covered the basic arithmetic operations, including addition and subtraction, and multiplication and division. The section also included working with fractions and decimals.

- **ADDITION AND SUBTRACTION:** Addition (getting a total of numbers) and subtraction (decreasing numbers by an amount) are covered in this section. They are the two simplest operations you can do with numbers.

- **MULTIPLICATION:** Multiplication allows you to perform repeated addition.

- **DIVISION:** Division is the process of splitting a number into parts or groups.

WORD PROBLEMS: This section of the guide went over how to do different types of word problems, the common wording you might encounter, and examples.

- **BASIC PROBLEMS:** These typically require one or two steps to complete.

- **PERCENT AND INTEREST:** These problems typically deal with money, such as adding interest back to an original amount.

- **RATIOS AND PROPORTIONS:** Ratios are used to compare two values, and proportions are ways of setting up and solving equations with ratios.

- **PROBLEMS OF MEASUREMENT:** These are the geometry type of problems, such as solving for perimeters and areas.

Takeaways

The word problems that are being presented in this section of the ASVAB are not very complex, and will typically only use basic math, arithmetic, and lower level algebra. If you don't know what you are doing, of course, they might prove to be too difficult to take on. Practicing this, knowing what types of questions you will encounter, and learning to do these basic math problems without the help of a calculator will help you immensely when you are going through this section of the test.

Here are a few final things you will want to think about before you mark down your questions:

- Does the answer you have reached seem like It is correct, based on its scope?
- Is the question that is being asked actually being answered? It can be very easy to accidentally get fooled by an irrelevant answer choice.
- Is the answer you have selected using the same measurement (units) as the problem is using?
- Can you work backwards from the answers supplied, if you can't work from the questions to the answer?

six

WORD KNOWLEDGE

Introduction

The word knowledge subtest is basically, at its core, a vocabulary test. Having a good vocabulary is going to be necessary if you want to move forward with a career in the military. Military operations and logistics require clear communication at all levels. This is one of the four subjects that comprises your AFQT score, and is one subjects that is most essential to your success and your future career with the military.

Qualification for many jobs within the military depends on your success on the word knowledge test as well. The following line scores utilize the word knowledge score:

Table 6.1. Line scores by military branch

BRANCH	LINE SCORE
Army	Clerical, general technical, skilled technical, operators and food, surveillance and communications
Marines	General technical
Navy/Coast Guard	Administrative, health, nuclear, general technical
Air Force	Administrative and general

There are two types of questions which will be asked in this subtest. The first will ask for a definition and the second will require you to determine the definition based on its context in a sentence.

EXAMPLES

Below is an example of the first type of Word Knowledge question:

<u>Word</u> means:

A. Definition 1
B. Definition 2
C. Definition 3
D. Definition 4

Below is an example of the second type of Word Knowledge question:

The man went into the <u>adjective</u> house.

A. Definition 1
B. Definition 2
C. Definition 3
D. Definition 4

One of the more unfortunate realities of the ASVAB is, however, that you usually won't be able to find the definition for a word you do not know through the use of context. It is hard to define a word, at times, based only on a single sentence. In the case of the first type of question, you will have no context at all. What most people learn as children, in order to assist them in learning new words, is to utilize context. That is something you will have to move past in order to have a lot of success on this test.

Study Information

The study of word knowledge is not as simple as you might think. There are, however, some things that you can study which will assist you in learning new words quickly. This guide will cover roots, suffixes, prefixes, basic parts of speech, and a few other things which can quickly assist you in learning large groups of new words quickly. There are two other things, which will not be covered here, which could greatly assist you in the learning process: learning German or learning Latin. Latin forms the basis for a huge number of words in English, and is evidenced greatly in many common roots of words. English itself if a Germanic language and learning German can assist you in your goal of learning more English as well.

Know Your Roots

Roots are the main part of a word. If you know common roots, you will be able to figure out what words mean (in most cases). The words formed from a given root do not have to be the same part of speech either. You can have the same root used in nouns, verbs, or even adjectives. Here are some of the most common root words you might encounter:

Table 6.2. Common root words

ROOT	DEFINITION	EXAMPLE
ast(er)	star	asteroid, astronomy
audi	hear	audience, audible
auto	self	automatic, autograph
bene	good	beneficent, benign
bio	life	biology, biorhythm
cap	take	capture
ced	yield	secede
chrono	time	chronometer, chronic
corp	body	corporeal
crac or crat	rule	autocrat
demo	people	democracy
dict	say	dictionary, dictation
duc	lead or make	ductile, produce
gen	give birth	generation, genetics
geo	earth	geography, geometry
grad	step	graduate
graph	write	graphical, autograph
ject	throw	eject
jur or jus	law	justice, jurisdiction
log or logue	thought	logic, logarithm
luc	light	lucidity
man	hand	manual
mand	order	remand
mis	send	transmission
mono	one	monotone
omni	all	omnivores
path	feel	pathology
phil	love	philanthropy
phon	sound	phonograph
port	carry	export
qui	quiet	quiet
scrib or script	write	scribe, transcript
sense or sent	feel	sentiment
tele	far away	telephone
terr	earth	terrace
uni	single	Unicode

vac	empty	vacant
vid	see	video
vis	see	vision

Prefixes and Suffixes

If you have a good knowledge of common prefixes and suffixes, then you will have a pretty good idea of how to figure out what a given word means, even if you have never encountered that word before. A prefix is a word part attached before the root (core part) of a word. A suffix is a word part attached to the end of a word.

Familiarizing yourself with prefixes and suffixes will help you to figure out the meanings of words by simply breaking them down into their component parts. Utilize the following sections and tables to help you kick-start your study of suffixes and prefixes.

Here are some of the most common suffixes you might run into on your test:

Table 6.3. Common suffixes

Suffix	Definition	Example
-able	capable of	actionable
-ian	belonging to	contrarian
-ile	relating to	rile
-ary	of or relating to	contrary
-ion	action	action
-ate	make	create
-ic	relating to	terrific
-ious	having the quality of	precious
-ity	state of being	piety
-ive	performing	live
-ism	practice of	syllogism
-ist	one who performs	psychiatrist
-ise	to cause/become	wise
-ize	to become	actualize
-ly	resembling	spindly
-less	without	windless
-y	state of	windy
-ry	state of	sundry
-ology	study of	geology
-ment	action/process	augment
-ship	position	relationship

-er	someone who	teacher
-dom	place	kingdom
-ify	make	edify

Prefixes will always come at the beginning of words. These can help you understand a word by acting as a modifier to the root word. Learning how language works, and how word parts come together to form whole words are a great way for you to go about learning language.

An example of a prefix would be *a-* (without the hyphen, of course). When used as a prefix, *a-* would mean *not*. For instance: *apathy* (without feeling). You can see, in this case, that the prefix fits right onto *pathy*, the root word. In this way, a new word is formed. The word itself, however, is just a combination of two word parts. Knowing those parts would allow you to understand this word whether you had seen it before or not.

Another example would be *anti*. Everyone knows this one. *Anti* means *against*. So *antithesis* would be the opposite of a given idea. *Antipathy* would be *against your feelings*. So on and so forth. Nothing too complicated. Prefixes are nothing to get worried about if you do not know them. Chances are, you know a lot of prefixes and simply don't realize that you do. They come easily over time once you start paying attention to them. The below table contains some of the most common prefixes you might encounter:

Table 6.4. Common prefixes

Prefix	Definition	Example
a-	without	amoral
an-	without	androgynous
ante-	before	antecedent
anti-	against	antigravity
auto-	self	autopilot
co-	with	cohabitation
de-	of	demure
dis-	not	disengage
en-	put into	engage
ex-	out of	extra
extra-	more than	extrasolar
hetero-	different	heterosexual
homo-	same	homophone
hyper-	over	hyperextend
il-	not	illegal
im-	not	immoral
circum-	around	circumnavigate
macro-	large	macroeconomics

micro-	small	microbiology
mono-	one	monorail
non-	without	nonissue
omni-	multiple	omnifocus
pre-	before	prefix
sub-	under	subscribe

Antonyms and Synonyms

Antonyms are words that mean either the opposite or almost the opposite meaning of the word in question. Synonyms are, interestingly enough, the antonyms of the word antonym. These are words that mean either the same thing or almost the same thing as the word in question.

On the ASVAB, you will be asked, at times, to find a word that most closely matches the definition of the given word (find the synonym of the given word). You might also be asked for find the word that is closest to the opposite of a given word (find the antonym of a given word). It would pay to, as you learn new words and look them up in the dictionary, closely look at any synonyms or antonyms that are listed by the definition. Online dictionaries usually list them. Print dictionaries do sometimes, but your mileage may vary depending on your specific dictionary.

Tips

The single biggest tip that will help you on this section of the test is also the one you can't do very quickly: Read a ton. The more you read, the more your vocabulary will naturally increase. Unfortunately, this is a long process, so the sooner you are able to start, the better. It will help if you read books and articles that provide you with a challenge, rather than ones that you find easy to read.

Learn your roots, prefixes, and suffixes. If you know them, then you will greatly expand your working vocabulary whether you like it or not. Knowing the most common examples of each of those will help you understand many words, whether you have heard them before or not, simply by using logic.

Be careful with words you have heard out loud but have never seen spelled before. *C* can sound like *S* or *K* at times. Everyone knows examples of these types of tongue twisters. If you hear a word out loud, try to make sure you know how to spell it as well, just in case it comes up on the test.

When you are reading, always push to learn new vocabulary words. Either try to figure them out through context (then confirm later) or simply look them up immediately. You can't grow by sticking with what you already know, so challenge yourself to learn more every time you can.

Anytime you hear a word you don't know, write it down and keep a list. Look the words up when you get a chance.

Use new words you find in a sentence when you know their meaning.

Try and make up example sentences that you can use the new word with. This will help to imprint it in your mind.

As strange as it may seem, studying other languages (and linguistics in general) will help you a lot with this. Many English words are derived from words in other languages (particularly German and Latin).

Spread your study out as much as you can. You will want to learn a new word every day, if possible. With the help of various vocabulary building sites and apps for smartphones, it should be no issue and can be done automatically in many cases.

Remember: If you know three of the four words in the answer box, you will automatically be able to rule them out (or rule one of them in), meaning you don't have to know the true definition (necessarily) to get a question correct.

Utilize other resources every chance you get to increase your vocabulary. Here are a few you can start using today:

- Dictionary.com
- Vocabulary.com
- Freevocabulary.com
- M-w.com
- Books such as SAT study guides, GRE study guides, or books in the For Dummies series dealing with vocabulary.
- Also, obviously, if you find a word you do not know, run a simple Google search.

GO ON

Practice Questions

1. What does <u>gouge</u> mean?

   A. decorate without taste

   B. completely enclose

   C. make a groove in

   D. convert text to computer code

2. Which of the following is closest to the term <u>refractory</u>?

   A. a grandmother who is sympathetic

   B. a defiant teenager

   C. a charismatic business man

   D. a dedicated police officer

3. Which of the following words most closely matches the definition of <u>supersede</u>?

   A. gain superpowers

   B. replace

   C. scare

   D. undulate

4. What does <u>stolid</u> mean?

   A. being down in the dumps

   B. showing excitement

   C. characterized by high humidity

   D. revealing very little sensibility or emotion

5. What does <u>placate</u> mean?

   A. destroy

   B. appease

   C. make worse

   D. hard

6. <u>Inadvertently</u> means:

   A. accidentally

   B. with contempt

   C. did alone

   D. immediately

7. "The informant <u>apprised</u> the officer of the situation." What does <u>apprised</u> mean in this context?

   A. told negatively

   B. provided with food

   C. got rid of

   D. made aware of

8. What does <u>efficacy</u> mean?

   A. having legal ability

   B. being judicious

   C. the power to cause a desired effect

   D. tolerance

9. What is an antonym of <u>diffidence</u>?

   A. confidence

   B. alarm

   C. aversion

   D. longing

10. What is the definition of <u>tractable</u>?

    A. reacting to suggestions quickly

    B. unhealthy

    C. softly

    D. in a strong way

11. <u>Lethargic</u> means:

    A. not reciprocated

    B. lacking activity or alertness

    C. having good fortune

    D. alone

12. The closest word to the definition of <u>pragmatic</u> is:

    A. precocious

    B. lonely

    C. practical

    D. conductive

13. To be deliberately <u>ambiguous</u> or unclear is to:

    A.   equalize

    B.   compare

    C.   dress well

    D.   equivocate

14. To be mixed with <u>impurities</u> is to be:

    A.   seamless

    B.   wasteful

    C.   adulterated

    D.   dirty

15. To <u>discredit</u> is to:

    A.   predict

    B.   reject as false

    C.   make warm

    D.   alert

Word Knowledge Answer Key

1.	C.	9.	A.
2.	B.	10.	A.
3.	B.	11.	B.
4.	D.	12.	C.
5.	B.	13.	D.
6.	A.	14.	C.
7.	D.	15.	B.
8.	C.		

Review and Takeaways

Learning the parts of words will help you immensely when you are attempting to expand your vocabulary. Take the practice questions and make not of any words you do not know (including words which are in the answer boxes that are not the correct answer. Of course, along with these practice questions, you should take the included practice tests as well. In addition, you should take any online or book-based ASVAB practice tests you can get your hands on. Don't forget to time yourself!

ROOT WORDS are main word parts which will help you determine what the full word meant. They are usually based in either Greek or Latin.

PREFIXES AND SUFFIXES se are word modifiers which can come either before (*pre*) or after (*suf*) the root word.

SYNONYMS are words that mean the same thing, and **ANTONYMS** are words that mean the opposite of each other.

seven

PARAGRAPH COMPREHENSION

Introduction

The purpose of the paragraph comprehension section of the ASVAB is to figure out how well you understand the things that you read. It also measures the level of your ability to retain information that you have read in passages. This is one of the four subtests on the ASVAB which has its score counted toward your AFQB score, so scoring a high number on this subtest is important to your future military career. There are fifteen questions on this test which are based on either fifteen distinct passages or a number of passages less than fifteen. Each passage will have a set of questions that is associated with it and you will generally have to pick the answer that best answers a question or fills in a statement to make it complete.

There are two skills which are used to help you understand the things that you read. One of them is the ability to understand what the passage you have read actually says. This is the literal reading of the passage. Some of the questions found on the test are going to ask you to determine what a passage means or to paraphrase something which has been said. To understand a paraphrase, you will have to understand what the original passage means, obviously. This also means you need to understand words in the context of the passage. Vocabulary will help with this.

The second skill you need is the ability to analyze the things that you read. This means you will have to go a bit deeper than the literal interpretation of the passage. You will be asked, at times, to draw conclusions based on information contained within the passage itself. This will often require you to figure out things which are stated indirectly (or not at all) in the content of the passage. Sometimes you will be asked about the tone of the passage or the mood that it evokes.

Reading Concepts

There are several types of reading concepts that you will likely encounter on this section of the test:

- main idea
- sequence of events
- rewording of facts
- stated facts
- mood and tone
- purpose
- technique
- conclusions

Main Idea

This type of question wants you to give a general statement about what the paragraph you have read means.

EXAMPLE

The economy is making a slow comeback. The housing boom of the 90s is not back, but it should be clear that the economy is improving from where it was just a few years ago. This is even in the face of fruit markets which have been slowing down, causing trouble for bankers and others as well.

What is the main idea of the prompt listed above?

A. The economy is not doing well.

B. The economy is on its way back.

C. Bankers are losing money.

D. The apple market isn't good.

The answer here is B. The thesis statement in the first sentence says the economy is on an upswing. Bankers may be losing money, the apply market (and other fruit markets) may not be doing well, but the economy in general is improving, making B the best choice in this situation.

Sequence of Events

These types of questions are asking about the order that events occur in the passage.

EXAMPLE

The walls were the first thing to fall. Once they had fallen, ranks quickly began to break and the soldiers prepared for a full retreat. Finally, there was the castle itself, which was the last to fall.

In what order did things happen?

A. Castle fell, then retreat, then the walls fell

B. Retreat, castle fell, walls fell

C. Walls fell, retreat, castle fell

D. Walls and castle fell followed by retreat

The answer is C. The events too place in a clear order. First the walls, then the retreat, then the castle fell.

Rewording of Facts

These types of questions will ask about facts in the text, but the answers will not have the exact same wording as the passage does. Instead, they will mean the same thing but the wording will be different.

EXAMPLE

Apple sales at the local market have been booming. 300 Granny Smith apples, 200 Fuji apples, and 150 Pink Lady apples have been sold in the last week alone. Those numbers are huge compared to what they were before the latest health food craze hit this part of the state. It is very good for apple farmers as well.

A. More Fuji apples were sold than anything else

B. More Pink Lady apples were sold than Fuji.

C. Granny Smith apples were the top seller last week.

D. People hate Granny Smith apples.

The answer is C. It is clear that they sold the most. This is a simple comparison problem, even without doing any math at all you can see that Granny Smith apples were the bestselling type of apple last week, which is another way of stating what is being said in the prompt.

Stated Facts

This type of question primarily relies on facts which are stated in the passage. You should avoid using any outside information for these questions. In addition, your answer will need to say everything that is in the passage with regard to the question itself. Often, you will want to find an answer that uses the exact wording that you found in the passage.

EXAMPLE

Ten boys went to the church function, along with five girls, six adults, and no children under two years old.

How many children under two years old went to the church function?

A. ten

B. two

C. three

D. zero

The correct answer here is D. No children under two years old went to the church function.

Mood and Tone

Mood and tone questions are about the emotions that are suggested by the content that you have read.

EXAMPLE

It was a beautiful day. The sun was shining and everything was going great. Blue skies and white clouds were everywhere and, best yet, my little cousin and I were at the park together eating cotton candy!

What emotions does this prompt elicit in the reader?

A. happiness

B. sadness

C. anger

D. loneliness

It is pretty clear that A, happiness, is the correct answer in this situation. Blue skies, cotton candy, the park? What could be happier than that?

Purpose

This type of question concerns itself with the purpose of the passage.

EXAMPLE

Welcome to the manual for your new blender! Here are a few things you need to know to get started using it! Plug in the AC adapter into the nearest wall outlet and then press the power button to turn the blender on.

What is the purpose of the above prompt

A. to explain how to use the new blender

B. to say what the blender is for

C. to get someone to buy the blender

D. to blend something

A. This is a simple one. Manuals typically serve the purpose explaining how to use a product. This type of question is going to be straightforward and relatively easy to answer.

Technique

Technique questions want you to identify techniques that form the basis of the structure of the passage you have read.

Conclusions

Conclusions questions are about indirect conclusions that you can infer from the text you have read.

Study Information

The general method of study for the paragraph comprehension portion of the test involves studying the individual type of questions that you will encounter in it. Obviously that sounds simpler than it actually is, but this section is, as the word comprehension, a matter of knowing what to prepare for rather than cramming. This section is akin to a "know it or don't know it" type of situation. The best thing to do is know the types of questions, learn the general tips, take practice questions and tests, and read voraciously. Doing those things will give you the best foundation for this subtest.

Main Ideas

The main idea section concerns itself with exactly what you think: the main idea. Main ideas are going to be general statements that sum up what a given passage is about. Information about the paragraph itself is usually specific and provides support for the main idea, which is often outlined in the very first sentence. Sometimes the main idea is referred to as a thesis statement.

The main idea is sometimes stated directly, but not always. If it is stated, it might be referred to as the topic sentence. Usually happening right at the beginning of a paragraph. Usually, the main idea or the topic sentence will not be found in the middle or the end of a passage, but it can be.

Usually, when you are asked to talk about the main idea, the right answer might be worded differently than the way it is stated in the paragraph. If the writer has chosen to not directly state the main idea, you need to figure out what it is by reviewing all of the information in the passage and determining what general point is being made. This is known as "inferring" the main idea.

The first thing you will want to do when you are trying to come up with the main idea for a given prompt is to look at the very first and the very last sentences of the passage. This does not mean you should neglect to read the rest of the passage, but those two sentences are generally the most important. Take a look at the following prompt, for example:

EXAMPLES

Local farmers are having trouble watering their crops due to the drought. Since last year, over ten thousand gallons of water have evaporated from the local fields. Combined with the lack of rain, local crops are dying off. Farmers might have reduced crop yields as a result.

What would be the main idea here?

The main idea is right there in the first sentence. *Local farmers are having trouble watering their crops due to the drought.* **The rest of the paragraph is only supporting evidence for that main point. You should keep in mind, however, that the main idea is not always going to be the one in the first sentence. Sometimes a paragraph is going to build up the main idea before stating it.**

Local gardens are beginning to blossom. Bees are buzzing near the flowers and flying around the trees. Kids are beginning to come outside instead of staying inside out of the cold. The sun is warming everything up. Spring is here!

The main idea in this case is stated in the final sentence, "spring is here". The rest of this paragraph is being used in order to

> **provide some context and supporting evidence for that main idea.**
>
> **Note: The main idea, again, is not always going to be stated outright. Sometimes it is going to be indirect or implied. One of the answers, however, should obviously be talking about the main idea.**

Writers do not always talk about only one specific point, however, so you need to be prepared to identify sub-points in your paragraph. Details are often included which will support the main idea and they will usually be helping you to prove whatever point the main idea is stating.

If you are having trouble coming up with the main idea, you might try attempting to paraphrase the prompt that you are reading. You can do that without even putting a lot of effort into it, simply do it in your head while you read. If you begin doing this with the things that you read now, you will find yourself improving very quickly.

Drawing Conclusions

These types of questions want you to draw conclusions based on what is in the passage. Often, the writer will not directly state the conclusions that you are meant to draw, but they will be indirectly stated. Usually, they are obvious. Take the pieces of information which are in the passage and then put them together to see what sorts of things are implied. The passage usually will not give you an answer to your question directly and the things which are stated directly in the passage are not conclusions. The conclusion is based on the relationships between the information you have been presented.

> ## EXAMPLE
>
> Most burglaries that occur in residential homes are because people neglected to lock their house up properly when they leave. The resulting crimes, thus, are crimes of opportunity.
>
> What sorts of conclusions can you draw here?
>
> > Well, there are a few, not all of them are correct, however. Here are some of the possible conclusions you could come up with:
> >
> > **Some crimes only occur because someone was neglectful of safety.**
> >
> > **Preventing opportunity will prevent crime.**

Most burglaries that occur in residential homes are because people neglected to lock their house up properly when they leave. The resulting crimes, thus, are crimes of opportunity.

Stated Facts

These questions are easily the simplest questions on this subtest. All they do is ask you for a fact which was stated in the passage. The wording will be the same. The exact same statement from the answer will be seen in the passage. Do not be tempted to use outside information which is not located inside of the text. The only information that you will be getting here is going to be inside the passage. Again: The information is ONLY about what is contained in the passage and nothing else.

EXAMPLE

Fuji apples were originally created by gardeners working in Japan in the 1930s. They are a type of hybrid apple which resulted from a combination of Red Delicious apples and Virginia Ralls Genet apples. They are large, round, and very sweet (especially when compared with some other types of apples).

Take a moment to look over this and find some of the stated facts that are in this paragraph. Once you have done that, take a crack at the following question.

The Fuji apple is an apple that is—

A. large, round, and very sweet

B. large, round, and yellow

C. a non-hybrid apple

D. grown only in Japan

The answer here is A. That is the only fact that is stated directly. It is stated in the very last sentence of the paragraph.

What are some other facts that are in this paragraph?

◆ Fuji apples are a hybrid of Red Delicious and Virginia Ralls Genet

◆ Fuji apples are sweet when compared to other apples

◆ Fuji apples originated in Japan

◆ Fuji apples were originally created in the 1930s

Mood and Tone

The mood and the tone of a given passage are a representation of the emotions that the content is trying to elicit in the reader. When you are faced with these types of questions, you will want to think about the type of language that is being used. Are they happy? Are they sad? You can usually tell immediately. If the skies are said to be dark and it is raining, the mood is likely sad or depressed. If the sun is out and the sky is bright, the mood will likely be happy. Think about how you might feel if you were suddenly dropped into the world of the passage. Whatever emotions you would likely be feeling at that time are the ones you will probably be feeling in the context of the passage itself.

Purpose

Questions about purpose will try and get at what the passage is intended for. What it is aiming to accomplish. Some writings are meant to provide the reader with information. Some are meant to provide instructions on how to do something. Some are persuasive and try to convince the person reading it of something. The purpose would be what the writing wants to do.

When trying to figure out what a passage is purporting to do, you should look at how the various sentences within it connect and relate with one another. If the passage is mostly evidence for the thesis statement, it is likely an argumentative passage. You should readily be able to tell if it is instructional or if it is meant to tell a story. You can practice this by doing it for each paragraph of the things that you read.

The conclusions that you draw about the meaning of a given prompt is what allows you to get to new ideas which are not directly stated in the text. All of the information which is being given to you by the author should be analyzed in order to find inferences that may be present in the text. You can use the example prompt from earlier in order to illustrate this idea:

EXAMPLE

Local farmers are having trouble watering their crops due to the drought. Since last year, over ten thousand gallons of water have evaporated from the local fields. Combined with the lack of rain, local crops are dying off. Farmers might have reduced crop yields as a result.

One thing you could infer from this text, for instance, is that the additional of rainfall would correct the problem that the farmers are having with their crops. The crops are dying because of the lack of water and the lack of rain. The addition of water/rain would fix this problem. Clearly this is not directly stated by the author, but all of the information required to come to that conclusion is right there in the text.

What are some other things that you might be able to infer from this passage? Here is a brief list of a couple:

The drought probably began last year.

The issue is a combination of heat and lack of rain ("evaporated" is the keyword).

Technique

Writing is always organized into some sort of structure. Various techniques are used to do this, and the technique questions on the ASVAB concern themselves with those techniques. They might ask you how a passage is structured. They might ask about keywords. Often, there are some tells that you can use to try and determine what sort of passage you are reading and these same words can be used to help determine the sequence of events and the purpose of the passage as well.

Words are the key to figuring out the technique being used. **NARRATIVE TECHNIQUE** uses words like *first, soon, then, next,* or *might* provide you with brief time frames of when events are happening. On the other hand, **DESCRIPTIVE TECHNIQUE** will use spatial descriptions of what is happening and might utilize the senses (sight, sound, taste, touch, smell). Also, look for words that relate things to each other in space like *on, next to, beside, under,* etc. **COMPARISON** is used as a technique with words like *similarly, like, same,* etc. Meanwhile, **CONTRAST** can also be done, using words like *as opposed to, on the other hand, but,* etc. If information is presented about why things are happening, this can be the technique of **CAUSE AND EFFECT**. Look for things like *since, because, resulting in, so,* etc.

In this section, you might also encounter some **LITERARY TECHNIQUES.** Some that you have heard of, some that you may not have heard of. Here is a brief rundown of a few of these types of techniques that you can learn:

Table 7.1. Commonly used literary techniques

TECHNIQUE	DEFINITION
Simile	A simile is a technique that is used to compare things with the usage of some sort of connecting word (than, so, as, like, etc.).
Metaphor	This is a technique which is meant to compare unrelated things through the use of rhetorical effect.
Ellipsis	the deliberate omission of certain words
Elision	omitting letters in speech. This is what leads to the colloquial speech that many people use when texting or chatting online.
Hyperbole	a form of exaggeration
Onomatopoeia	using a word to imitate a real sound (such as *boom* for an explosion)

Obviously there are many techniques, but this should give you a good enough basis for what you are likely to find on the ASVAB.

Sequence of Events

Questions about the sequence of events are exactly how they sound. They are asking about what order the events of the prompt are occurring in. You will want to spend your time looking at the prompt and picking out words that talk about time. These could be words such as *before, then*, *next*, *finally*, etc. These are the types of words which indicate what events happened and when they happened. This is a relatively simple thing to do and these types of questions are usually not too complicated.

Reworded Facts

Reworded facts are pretty easy to understand. All you need to do is look for facts that have been stated in the prompt to answer the questions. Usually, you will have to pick an answer which means the exact same thing as something which has been stated in the passage itself, though the words will not always be the same.

Tips

Don't bother reading the instructions in this section of the test. Not only will they be completely obvious based on the questions, but you already know what you will have to do because you have read through this guide. All the instructions will do is tell you something that you already know and waste some of your valuable time.

There are two ways to go about reading, and you should do both. Quickly skim the passage to get the main idea, then skim the questions. Once that is done, run back through the passage based on the information that is being asked by the questions. Sometimes close reading is not necessary.

Don't spend a lot of time on individual questions. If you do not know one, go to the next. You can always go back and answer later (on the pen and paper). With less than a minute per question, you simply can't afford to spend your extra time with questions that you don't know the answer to.

As always with multiple choice tests, narrow down your options as much as possible before choosing something to go with. This will help you if you wind up having to make an educated guess to get the answer.

Answers have to be based on the information in the given passage only. If it isn't there, then don't take it into consideration. This can be particularly hard to do if you know information that has not been provided. Fight the urge to use that information.

To quickly improve on this section of the ASVAB, read more. Read as much as you can. Read things which are not easy for you to read. Expand your vocabulary if possible. Read newspapers, magazines, and books. When you read, make sure you have both a direct and an indirect understanding of the text. In addition, practice your recall when you can by writing down a list of things you remember from the text every time you finish your reading.

Be confident in the things that you are reading. Most people begin to falter on this section of the ASVAB because they constantly second-guess the answers that they have come up with or they believe that they are unable to come up with the correct answers on their own. Chances are, you are capable of doing coming up with the answers on your own. Chances are, in fact, the first answer you believe is correct is the correct answer.

Combine your study of this subtest information with the information contained in the word knowledge section. Both of these subtests complement each other.

If you spot the word *never* in one of the answers (or any word like it; *always*, *forever*, etc.), don't pick that answer. It is very unlikely any answer giving an absolute is going to be the correct one.

Practice Questions

Prompt One

California is in the middle of a long drought. It is said that it would take a full two years for them to fill up the aquifers which have been depleted as a result of this drought. Even so, they are quickly coming up with plans to fight against the lack of water, including building new desalination plants. They have not, however, bothered to spend time cutting back on the amount of water people are allowed to use to water their lawns or, worse yet, golf courses.

1. What is the problem that California is currently undergoing, according to the prompt above?

    A. drought

    B. too much rain

    C. water being stolen

    D. salt water

2. Which of the following figures of speech is being used heavily in the prompt?

    A. onomatopoeia

    B. exaggeration

    C. metaphor

    D. none of the above

3. Why are the desalination plants being built?

    A. to fix the drought

    B. to get more water

    C. to water golf courses

    D. to help people move to higher ground

4. Which of the following is the best description of the drought?

    A. Golf courses are drying out.

    B. Salination plants exist.

    C. The aquifer will take two years to fill.

    D. It is raining entirely too much.

Prompt Two

The United States of America has outlined customs and rules about how the flag of the country can be shown and displayed in a respectful and proper way. For one thing, the flag may only be shown during daylight hours (sunrise to sunset). It can be shown at night only if it can be lit up so it is seen clearly even in the darkness. Everywhere that voting is taking place needs to have flags clearly visible. The flag should never touch the floor or the ground. Flags must not be burned or damaged in any way. Flags should also never be utilized for any sort of marketing.

5. John has his flag out at night under a light, it is located on the roof of his car dealership next to a large sign advertising his prices. What did John do wrong?

    A. flag out at night

    B. the flag being used for marketing

    C. flag not visible

    D. flag touching the ground

6. When can the flag be shown at night?

    A. when it is lit

    B. it can't

    C. when it is on the ground

    D. when it is used in images

7. Which situation is allowed?

    A. flag displayed at night

    B. flag displayed during the day

    C. flag on the ground

    D. flag in advertising

8. Which of the following statements is accurate?

   A.   Nobody has ever disrespected the flag.

   B.   Rules and customs are made to be broken.

   C.   The United States places importance on the flag.

   D.   The United States does not place importance on the flag.

Prompt Three

The unspoken truth of the situation was that there really was something going on in the house. Even with the trees brushing on the windows and the fact that there was a graveyard behind the house, Janet had never really believed that it was haunted. It was, however, becoming very hard to argue with the fact that there was something odd going on, given the fact that she was currently looking at a mirror floating five feet above the ground.

9. What is the mood and tone of this passage?

   A.   matter of fact

   B.   happy

   C.   sad

   D.   depressed

10. What made Janet begin to believe that the house might be haunted?

    A.   the ghost

    B.   graveyard out back

    C.   trees brushing the windows

    D.   floating mirror

Prompt Four

Professional golf players are usually able to hit the ball a bit faster in modern times than they ever could before. There are a couple of reasons for this. For one thing, they are fitter and better conditioned. They have higher strength levels. Beyond that, there is also the fact that modern golfing equipment utilizes different materials?

11. What is one possible reason pro golf players today can hit farther?

    A.   They weight train.

    B.   They use steroids.

    C.   They are made of iron.

    D.   They are not old.

12. What is one inference you can draw from this?

    A.   Golf players used to be more conditioned.

    B.   Materials are worse now than before.

    C.   Golf players used to be less conditioned.

    D.   Gravity is different, allowing for faster balls.

Prompt Five

I took a seat at the end of the hearthstone opposite that towards which my landlord advanced, and filled up an interval of silence by attempting to caress the canine mother, who had left her nursery, and was sneaking wolfishly to the back of my legs, her lip curled up, and her white teeth watering for a snatch. My caress provoked a long, guttural gnarl.

Wuthering Heights by Emily Bronte

13. What response did the dog have to being touched?

    A.   curling up

    B.   biting her lip

    C.   a snarl

    D.   sneaking around

14. Why is the narrator caressing the dog?

    A.   to get a snack

    B.   to fill the silence

    C.   so the dog would care for him

    D.   because it came in the room

Prompt Six

Her eyes were the moon reflected on a vast ocean of unimagined depth.

15. This is an example of:

    A.   onomatopoeia

    B.   exaggeration

    C.   simile

    D.   metaphor

GO ON

Paragraph Comprehension

Answer Key

1.	A.	9.	A.
2.	D.	10.	D.
3.	B.	11.	A.
4.	C.	12.	C.
5.	B.	13.	C.
6.	A.	14.	B.
7.	B.	15.	D.
8.	C.		

Review

The paragraph comprehension subtest is one of the most important tests on the ASVAB. It is one of the ones that figures into your main score on the test. You need to be able to understand the things that the passage is actually saying to you and, beyond even that, you need to be able to pick out and interpret specific facts and techniques which have been utilized in the prompt. This means you will have to be able to draw conclusions from the prompt, pick out main ideas and thesis statements, and analyze the techniques, purpose, and information which is contained within it.

The **MAIN IDEAS** section explains how to find the main idea in a prompts and what the prompt is supposed to be about. It also discusses thesis statements.

The **DRAWING CONCLUSIONS** section is about how to draw conclusions based on the prompt.

The **STATED FACTS** section deals with questions and tactics for determining facts stated inside of the prompt.

The **MOOD AND TONE** section will ask how the prompt seems to be written. How it makes you feel.

PURPOSE will have you answer why the prompt was written.

TECHNIQUES include various writing techniques and styles.

The **SEQUENCE AND EVENTS** section will have you recall the order that the things happen inside of the prompt.

REWORDED FACTS are stated inside the prompt itself, but these will typically word them differently. The purpose is the sa.me, but the way things are stated it a little bit different.

Takeaways

The more you read and the more you think about what you are reading, the better you will do on this. With only thirteen minutes to answer the fifteen questions, you will need to be on your toes. One thing you will want to keep in mind, however, is that you need to read through the prompts as quickly as possible and then read the questions. Once that has been done, go back through the prompt and pick out any additional information that is necessary to answer them.

eight

MATHEMATICS KNOWLEDGE

Introduction

The purpose of the MATHEMATICS KNOWLEDGE section on the test is to make sure you fully understand the concepts that are important in high school mathematics courses. This includes information about basic operations, order of operations, algebra, and geometry. It also includes working with fractions, which can prove difficult for many people.

Number Theory

NUMBER THEORY is the study of the properties of whole numbers (0, 1, 2, …) and also integers, which are whole numbers plus their negative counterparts. Negative numbers can be thought of as inverses or opposites of whole numbers. Integers, like whole numbers, can be written without the use of a fractional part.

Prime Numbers

Prime numbers are numbers that have only two factors, 1 and itself. Examples would include 2, 3, 5, 7, 11, 13, 17, 19, and 23, as well as many others. When you are attempting to figure out if a number is a prime number or not, all you need to do is figure out whether other numbers, besides 1 and the number itself, will divide evenly into it.

EXAMPLE

Is 66 a prime number?

66 is not going to be considered a prime number. It can be divided as 2 and 33, as 11 and 6, or as 1 and 66.

It is important to note that prime numbers are usually going to be odd, with the exception of the number 2. Even numbers clearly can be divided by 2, so it won't work for larger numbers.

Mean (Average), Standard Deviation, Median, and Mode

You may be asked to find the MEAN, STANDARD DEVIATION, MEDIAN, or MODE of a set of numbers. The mean (average) of numbers is obtained by adding up all the numbers and then dividing by the number of numbers that you added up. The standard deviation is a measurement (which you probably won't have to compute!) of how far apart the numbers are from the mean or average. The median is obtained by ordering the numbers and picking the middle number, or averaging the two middle numbers. The mode is obtained by finding the number repeated the most number of times (if there is a number that repeats).

Multiples

MULTIPLES of numbers are what results from multiplying whole numbers by other numbers. For example, multiples of 7 are 7, 14, 21, and so on.

Common multiples of two numbers are the numbers that are multiples of both. If you were looking at, for example, 4 and 8, then 16 would be a common multiple. 8 would also be a common multiple of the two numbers.

The least common multiple is the smallest common multiple that two given numbers share. The fastest way to find this is to just write out the first few multiples for both numbers that you have available and then figure out which one is the smallest. This is not a particularly difficult task, but you will have to understand how multiples work in order to do it properly. For example, the least common multiple of 4 and 8 is 8.

Factors

A FACTOR is a number that goes evenly into another number with no remainder. So think of factors as all the numbers you can multiply together to get another number. Here is an example:

EXAMPLE

What are the factors of 24?

1 and 24, 2 and 12, 3 and 8, 4 and 6, since all of these numbers go into 24 exactly (without a remainder).

It is not necessary to write them like that, but it will help you keep them in mind if you keep the factors together with their counterparts. Another way to write these is 1, 2, 3, 4, 6, 8, 12, and 24.

If you want to figure out whether a number is a factor of another one or not, just divide the number by the potential factor and see if the result is a whole number. Every number will have at least two factors, 1 and the number itself. For example, the two factors of 2 are 1 and 2.

Numbers that are the factors of more than one whole number are known as COMMON FACTORS. For example, 6 would be a common factor of both 12 and 24 (because it can be multiplied by 2 to reach 12 and 4 to reach 24). The largest factor that goes into two numbers is called the greatest common factor.

Exponents

EXPONENTS and exponential notation are used to help simply expressions, particularly when factors are repeated multiple times. Exponents are written as superscripts above the number that has the exponent. The number that has the exponent is called the base, so 8 is the base and 2 is the exponent in this example:

$$8 \times 8 = 8^2$$

Think of exponents as shorthand that is used to help keep the math straight and stop it from becoming too confusing. This can be seen below:

$$5 \times 5 \times 5 \times 5 \times 5 \times 5 = 5^6$$

Writing the number as an exponent makes it much easier to see what is happening and will simplify equations for you. Another way to say what is happening above is "five to the sixth power" or "five to the sixth". Here is another example where we are simplifying with an exponent:

$$2 + 2 \times 5 \times 5 \times 6 \times 7 + 8 = 2 + 2 \times 5^2 \times 6 \times 7 + 8$$

Does it simplify that equation a lot? Not in this case, but it does make it a little bit simpler to read.

Exponents are also used to denote cumbersome numbers in scientific notation; for example, $54,000 = 5.4 \times 10^4$ and $.0043 = 4.3 \times 10^{-3}$.

Let's say a few words about fractional exponents and negative exponents. In an exponent that is a fraction, the number on the top acts just like an exponent, but the number on the bottom designates a "root" (see next section), which means a number would have to multiplied that many times to get to that number. For example, $8^{\frac{2}{3}} = (\sqrt[3]{8^2}) = (\sqrt[3]{8})^2 = 2 \times 2 = 4$. With negative exponents, you have to take the reciprocal of the base and make the exponent positive. For example, $2^{-3} = \left(\frac{1}{2}\right)^3 = \frac{1}{8}$.

Also note that you can use exponents for questions dealing with combinations of letters or numbers. For example, if you were asked how

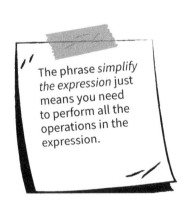

The phrase *simplify the expression* just means you need to perform all the operations in the expression.

many different numbers on a license plate there could be if the first three digits were letters, and the last 4 were numbers 0 – 9, you'd have $26^3 \times 10^4$, since there are 3 places for 26 letters and 4 places for 10 numbers.

Square Roots

The term **SQUARE ROOT** is used to describe a number that can be squared to equal the number provided.

The radical sign ($\sqrt{}$) is used to show square roots. For example, $\sqrt{9}$ = 3, since 3^2 = 9.

Some numbers will have very clean whole numbers for their roots. These are known as perfect squares. Here is a table that shows common perfect squares:

Table 8.1. Perfect squares

NUMBER	PERFECT SQUARE	SQUARE ROOT
1	1	$\sqrt{1}$
2	4	$\sqrt{4}$
3	9	$\sqrt{9}$
4	16	$\sqrt{16}$
5	25	$\sqrt{25}$
6	36	$\sqrt{36}$
7	49	$\sqrt{49}$
8	64	$\sqrt{64}$
9	81	$\sqrt{81}$
10	100	$\sqrt{100}$

Note that if we wanted the number that is multiplied 3 times to get a certain number, this is the cube root, so $\sqrt[3]{8}$ = 2. We can use this same notation for 4th roots, and so on.

Order of Operations

The order of operations is the way that multiple operations need to be done in order to reach the correct answer. Some operations take precedence over others; mathematical operations don't also go from left to right, like reading does.

Here is the basic order of operations:

1. First take care of any operations that are within a grouping symbol such as parentheses () or brackets [].
2. Next handle the roots and the exponents.
3. Next handle the multiplication and division in the same order that they appear (left to right).

4. Finally, handle the addition and subtraction (again, moving from left to right).

For an example of why this is important, consider the following:

Solve: $2 + 2 \times 2$

If you ignore the order of operations, what do you get?

$2 + 2 = 4 \times 2 = 8$

Now if you were to follow the order of operations, what happens?

$2 \times 2 = 4 + 2 = 6$

We get two different answers, but only the second is correct, since we need to perform multiplication before addition. It is important to eliminate any ambiguous statements in equations, because precision is key. That is why the order of operations is very important.

There is an acronym that can be used to help you remember the order of operations; **PEMDAS**:

- **P**arentheses
- **E**xponents
- **M**ultiplication/**D**ivision
- **A**ddition/**S**ubtraction

If you are using PEMDAS, you need to remember that multiplication and division have to be completed as one step from left to right, and the same with addition and subtraction.

Working with Integers

INTEGERS is a set that includes all whole numbers and the negatives (opposites) of those numbers as well. For example, integers are:

$$\ldots -3, -2, -1, 0, 1, 2, 3 \ldots$$

Think of adding negative numbers as the same as subtracting positive numbers. For example,

$$6 + -5 = 6 - 5 = 1$$

ADDITION AND SUBTRACTION WITH POSITIVES AND NEGATIVES

There are two situations you will encounter when you have to add integers with negative sign(s). Are the numbers the same sign or are the numbers opposite signs? If the numbers are the same, then you can just add them like you normally would, and then add a negative sign to the answer if the two signs were negative to start out with. Here are a few example situations:

$$3 + 3 = 6$$

$$-3 + -3 = -6$$

So how do you handle a situation when the two numbers have different signs? Ignore the signs, subtract the smaller from the larger, and then use the sign of the greater of the two numbers. Here are a few examples:

$$3 + -4 = -1: 4 \text{ is greater than 3, so use the sign before the 4}$$
$$\text{(negative) in the answer}$$

$$-5 + 4 = 1: 5 \text{ is greater than 4, so use sign before the 5 (positive) in}$$
$$\text{the answer}$$

If you are subtracting, and have two negatives together, change them into one positive:

$$2 - -2 = 2 + 2 = 4$$

In all, the best way to handle subtraction with negative numbers is to just turn the problem into an addition problem.

You may also need to know that the ABSOLUTE VALUE (signified by | |) is the positive equivalent of positive and negative numbers. So $|-4| = 4$ and $|4| = 4$. So a number and the absolute value of its negative counterpart are equal.

Multiplication and Division with Positives and Negatives

Again, the best way to handle this is to ignore the signs and then multiply or divide like you normally would. To figure out what sign the final product will have, you simply have to figure out how many negatives are in the numbers you worked with. If the number of negatives is even, the result will be a positive number. If the number of negatives is odd, the result will be a negative number.

For example:

$$2 \times 2 \times -2 = -8: \text{one negative; odd number of negatives, so}$$
$$\text{negative}$$

$$-2 \times -2 \times 2 = 8: \text{two negatives; even number of negatives, so}$$
$$\text{positive}$$

$$-2 \times -2 \times -2 = -8: \text{three negatives; odd number of negatives, so}$$
$$\text{negative}$$

Exponents of Negative Numbers

Exponents with negative numbers are relatively simple: If the number with the exponent (the base) is negative and the exponent is even, the result will be positive. If the number with the exponent is negative and the exponent is odd, the result will be negative. This is because, for example, $-4 \times -4 \times -4 \times -4 = 256$, but $-4 \times -4 \times -4 = -64$.

But we have to be careful with negative bases and even exponents. For even exponents, if the base is negative and in parentheses (or is a variable that you're putting in a number in for), the result is positive. For even exponents, if the negative sign is outside the parentheses, you have to raise the base to the exponent first, and then apply the negative (change the sign).

Here are some examples:

$$(-4)^2 = 16$$
$$-4^2 = -(4^2) = -16$$
$$(-3)^3 = -27$$
$$-3^3 = -(3^2) = -27$$

Evaluate x^2, where $x = -2$: $(-2)^2 = 4$

Evaluate $-x^4$, where $x = -2$: $-(-2)^4 = -16$

And again, remember that exponents are just another way of writing out multiplication.

Working with Fractions

Check the arithmetic review section of this guide for information on how to convert fractions back and forth. This section will cover different types of fractions and how to perform operations on fractions.

Equivalent Fractions

Equivalent fractions are fractions that are equal. One of the most common things that you will do when you are working with fractions is to simplify them. Another way to state this is to "reduce" fractions. All this means is writing the fraction in the smallest equivalent fraction you can (smallest top and bottom). For example, $\frac{5}{10}$ and $\frac{2}{4}$ can both be simplified to $\frac{1}{2}$. So, $\frac{5}{10}$, $\frac{2}{4}$, and $\frac{1}{2}$ are all equivalent fractions.

Here are some things to keep in mind when you are trying to simplify fractions:

- You need to find a number that can evenly divide into the top and the bottom number of the fraction that you are simplifying. After that, you can do the actual division.
- Once the division is finished, check to make sure your fraction cannot be further simplified. It is easy to make this mistake; even if you have found an equivalent fraction, it may be completely simplified, and your answer could be wrong.
- You can use simplification to reduce fractions to lower terms by dividing the top and bottom by the same number. You can also, perhaps more importantly, raise fractions to higher terms if you multiply both the top and the bottom numbers by the same number. This is very important in the addition and subtraction of fractions.

Addition and Subtraction

To add and subtract fractions, you need to understand two terms:

- **NUMERATOR:** The number on the top of a fraction.
- **DENOMINATOR:** The number on the bottom of a fraction.

If you have two numbers that have the same denominator, you will have what is known as a common denominator. You can really only add or subtract fractions that have a common denominator, so if you do not have one, you need to make one, using equivalent fractions.

Here are the steps for adding and subtracting fractions:

1. If the fractions have a common denominator, then proceed as usual. If not, then reduce or raise one or both fractions until you have a common denominator.
2. Add or subtract the numerators as you would any number, ignoring the denominator.
3. Place the resulting sum (or difference) on top of the common denominator as the new numerator.
4. Simplify the new fraction as much as possible.

Step 1 is involved with coming up with what is known as the least common denominator. This is the smallest number that all fractions have as a common denominator. The process of doing this is the same as the process for finding the least common multiple.

You can turn mixed numbers into improper fractions before adding, for example:

$$2\frac{1}{2} + 3\frac{3}{4} = \frac{5}{2} + \frac{15}{4} = \frac{10}{4} + \frac{15}{4} = \frac{25}{4} = 6\frac{1}{4}$$

You can also add the whole number parts, then add the fractional parts, and then add the two together. This will save you a lot of trouble and extra steps. It is just as correct as any other method of solving the problem and, of course, remember that you are not being tested on how you did it, just that it was done. Here is the same problem:

$$2\frac{1}{2} + 3\frac{3}{4} = 2 + \frac{1}{2} + 3 + \frac{3}{4} = 2 + 3 + \frac{1}{2} + \frac{3}{4} = 2 + 3 + \frac{2}{4} + \frac{3}{4} = 5\frac{5}{4} = 6\frac{1}{4}$$

You do have to be careful subtracting mixed numbers, since you may have to borrow:

$$2\tfrac{1}{4} - 1\tfrac{1}{2} = 2\tfrac{1}{4} - 1\tfrac{2}{4} = 1\tfrac{5}{4} - 1\tfrac{2}{4} = \tfrac{3}{4}$$

A simpler way may be to turn mixed fractions into improper fractions:

$$2\tfrac{1}{4} - 1\tfrac{1}{2} = \tfrac{9}{4} - \tfrac{3}{2} = \tfrac{9}{4} - \tfrac{6}{4} = \tfrac{3}{4}$$

Multiplication and Division

Multiplication and division of fractions is actually simpler than adding or subtracting them. When you have to multiply fractions together, you'll first want to turn any mixed fractions into improper fractions. Then you just multiply the numerators to get the new numerator and then multiply the denominators to get the new denominator. Once that is done, you can go ahead and simplify. For example:

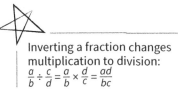

Inverting a fraction changes multiplication to division:
$$\tfrac{a}{b} \div \tfrac{c}{d} = \tfrac{a}{b} \times \tfrac{d}{c} = \tfrac{ad}{bc}$$

$$2\tfrac{1}{5} \times \tfrac{3}{7} = \tfrac{11}{5} \times \tfrac{3}{7} = \tfrac{11 \times 3}{5 \times 7} = \tfrac{33}{35}$$

There is no way to simplify $\tfrac{33}{35}$, so this is the final answer.

Division is a little bit more complicated, but ultimately it is the same procedure. First, you have to find the reciprocal of the second fraction. A reciprocal is a fraction that is "flipped": its numerator and denominator are switched. Once that is done, you will multiply the fractions as usual. Here is an example:

$$\tfrac{5}{6} \div \tfrac{2}{3} = \tfrac{5}{6} \times \tfrac{3}{2} = \tfrac{5 \times 3}{6 \times 2} = \tfrac{15}{12} = \tfrac{5}{4}$$

As you can see, the second fraction $\left(\tfrac{2}{3}\right)$ simply flips to $\tfrac{3}{2}$. The common way this is explained is to "flip and multiply". That is as good an explanation as any, and is certainly easier to remember. Again, you are being tested on your ability to do the math here, not to know the jargon.

Algebra

ALGEBRA is a method of generalizing expressions involved in arithmetic. You will be able to explain how groups of things are handled all the time. This is useful for times when you have a certain function that you need to do over and over again. In algebra, you typically have numbers as well as symbols (usually letters) called VARIABLES that can be used to stand for certain numbers. Typically, variables are useful when dealing with word problems, since algebra makes it easier to solve for an "unknown."

Evaluating Numbers

Numbers that are assigned a definite value (like the number "1") are constants. When symbols (variables) are used to stand for numbers, they can typically take on any number. If you saw the equation: $2x = 6$, x would be the variable here, and in this case it would equal 3.

A few things to keep in mind:

- You can make a variable anything you want. X, y, Z, a, A, b, and so on are examples. They are usually italicized, to distinguish them from numbers.

- Both sides of the = sign are, obviously, =. So you can add or subtract or multiply or divide anything on both sides (constants or variables or both), as long as you do it to both sides. This is how you solve equations. With inequalities, however, you must change the sign if you multiply or divide by a negative number.

- Many times, you must distribute either variables or constants through to get rid of parentheses. For example, if you have $3(x - 9)$, you'll want "push through" the 3 to make it $3x - 27$.

- Once you have figured out what a variable is, plug it into the original equation to make sure everything is still equal.

- You may be asked to evaluate a function for a certain value in the variable. Just plug in that value everywhere the variable is. For example, the value of $f(3)$ in the function $f(x) = 3x + 2$ is $3(3) + 2 = 11$.

Equations

Equations are expressions that have an equal sign, such as $2 + 2 = 4$. Equations will usually have a variable, and will always be true or false. $2 + 2 = 1$ is false. $2 + 2 = 4$ is true. When you are solving for variables, only one answer for each variable, usually, will make the expression true. That is the number you must find.

Basically, to do this, you will just rewrite the equation in more and more simple terms until you have the solution to it. Ideally, you want this to be x (or whatever variable) = a number.

$$x = \#$$

Again, you can do anything to an equation as long as you do the same thing to both sides. This is how you solve equations.

EXAMPLE

Solve $3(x - 2) = -7 + 10$.

First, simplify anything you can on one side (the −7 and 10) and also get rid of any parentheses by "pushing through" (the "3" on the left hand side):

$3x - 6 = 3$

Next, add 6 to both sides to get $3x$ by itself:

$3x - 6 + 6 = 3 + 6$

$3x = 9$

> Now, simplify this by dividing both sides of the equation by 3. Note: You have to divide the entire side by 3, not just one part of it:
>
> $$\frac{3x}{3} = \frac{9}{3}$$
>
> $$x = 3$$
>
> Now, we have x all by itself on one side; the equation is now solved.

Note that if we have an inequality instead of an equation, we have to be careful. We solve an inequality the same way, but if we multiply or divide by a negative number, we have to switch the sign.

EXAMPLE

Solve $-3x - 2 \geq 4$.

$$-3x - 2 \geq 4$$

$$-3x - 2 + 2 \geq 4 + 2$$

$$-3x \geq 6$$

$$\frac{-3x}{-3} \leq \frac{6}{-3}$$

$$x \leq -2$$

One more thing to mention here: If you are given an algebraic equation that is has 0 on one side and factors with variables, set each factor to 0 to solve for the variable. For example:

$$\text{Solve for } x: (2x + 6)(3x - 15) = 0$$

$$2x + 6 = 0 \text{ and } 3x - 15 = 0; x = -3, 5$$

An algebraic function like the one above is a polynomial of degree 2, which is a quadratic. To get the degree of a polynomial, add up all the exponents in each term, and the degree is the sum of exponents in the highest term. For example, for $4x^3y^5 + 8x^8y + 4x + 5$, the degree is 9, since in the second term, which has the largest sum of exponents, the sum of the exponents is $8 + 1 = 9$.

Word Problems

WORD PROBLEMS often utilize algebraic principles in their text. It is important to know how to properly assign a variable inside of a word problem. Pay attention to the words used: x *equals* or *a number equals*, x *is less than*, *2 is added to* x, are common wordings. A lot of times, you can just "translate" the English right into the math; for example, *a number added to 5* would be $x + 5$. Be careful though; *3 less than a number* would be $x - 3$ (do the math with simple numbers to figure out wording).

If you are not given a specific variable but you need one, just call it whatever you want. x is the simplest variable you can use to do this, and is also one of the least confusing when you start working with more complex algebraic principles.

Here is a type of word problem you might encounter with ratios: A class of 140 has sophomores, juniors, and seniors in a ratio of 4:2:1. How many juniors are in the class? To solve this, you can set it up like this: $4x + 2x + 1x = 140$; $7x = 140$; $x = 20$. Since there are $2x$ juniors, there are $2 \times 20 = 40$ juniors in the class.

There is another type of algebra problem that you might encounter, and it involves "work". You can remember this equation:

$$\frac{\text{Time to do a job together}}{\text{Time to do a job alone}} + \frac{\text{Time to do a job together}}{\text{Time to do a job alone}} = 1.$$

For example, if we had a problem like "If Mark can do a job in 4 hours and John can do it in 3 hours, how long would it take for the two to do the job working together?," we'd have $\frac{x}{4} + \frac{x}{3} = 1$, to get $x = \frac{12}{7}$.

Also, always remember that Distance = Rate × Time.

Multiplying Exponents

Multiplying variables with exponents is simple, as long as you have the same number with the exponent (the "base"). Just keep the same base and add the exponents together to get the new exponent. Keep in mind how exponents work as well; that is important.

Here is an example of exponents with variables:

$$x \times x \times x = x^3$$

Here is an example showing the multiplication of exponents; you can see why you add exponents:

$$x^2 \times x^3 = x^{2+3} = x^5, \text{ since } (x \times x) \times (x \times x \times x) = x^5$$

When raising something with an exponent in it to another exponent, you multiply exponents. And when dividing exponents, you do the opposite of addition, and subtract exponents with the same base. You may get a negative exponent; if you do, you can put it on the other side of the "division sign" and make it positive. For example:

$$\frac{(2x^2y)^3}{x^4y^5} = \frac{8x^6y^3}{x^4y^5} = 8x^{(6-4)}y^{(3-5)} = 8x^2y^{-2} = \frac{8x^2}{y^2}$$

Polynomials

POLYNOMIALS are just a list of algebraic terms that can contain variables, constants, and exponents, but can never have any term that is divided by a variable.

To combine like terms, you can add or subtract terms with the exact same variables. For example: $3x^2 + 2xy - 4 + 4x^2 - xy = 7x^2 + xy - 4$.

Note that if you have to multiply two two-term expressions to get a polynomial, use the FOIL (First, Outer, Inner, Last) method. For example:

$$(3x + 2)(x - 2) = (3x)(x) + (3x)(-2) + 2x - 4 = 3x^2 - 6x + 2x - 4 =$$
$$3x^2 - 4x - 4$$

If you have the same two-term expressions but with opposite signs (difference of squares), the middle terms (Outer and Inner) will cancel out:

$$(x + 2)(x - 2) = (x)(x) + (x)(-2) + 2x - 4 = x^2 - 4$$

Factoring

FACTORING is the process of breaking one quantity down into the product of some other quantity (or quantities). When you learned to use distribution to multiply out variables ("push through"), you are learning to do the opposite of factoring. When you factor, you are removing parts of an expression in order to turn it into factors. Basically, you do this by removing the single largest common single term (monomial) that is a factor; this is the greatest common factor.

EXAMPLE

Factor the following: $3x^2 - 9x$

Basically, you will pull out the largest common denominator between the two of them. In this case, it is $3x$. So you will be left with the following:

$3x (x - 3)$

To check this, "push through" the $3x$ to both the x and the 3, with a "−" sign in the middle; you get the original! It is also possible to do this with expressions that have more than two terms. Usually, these will be trinomials, with 3 terms. These will always end up being in the form of (x plus or minus something) times (x plus or minus something). What the second number is and what the sign is for the problem is determined by the original problem (we saw this above in the Exponents – Multiplication section).

Also, if you have a difference of squares, as we saw above, you can factor as the following:

$$(a^2 - b^2) = (a + b)(a - b)$$

SIMPLIFICATION

You can SIMPLIFY algebraic expressions the exact same way that you would fractions. Basically, you cake out the common factors. They just cancel out. You will want to multiply everything out if you cannot find a factor and then simplify. Here is an example:

$$\frac{(x + 4)^2}{x + 4} = \frac{(x + 4)(x + 4)}{x + 4} = x + 4$$

The other thing to remember is if you have to add fractions with variables, you may have to find a common denominator. For example:

$$\frac{x - 2}{x} + \frac{x - 2}{x + 2} = \frac{(x - 2)(x + 2)}{x(x + 2)} + \frac{x(x - 2)}{x(x + 2)} = \frac{x^2 - 4 + x^2 - 2x}{x(x + 2)} = \frac{(2x^2 - 2x - 4)}{x(x + 2)}$$

Geometry

The GEOMETRY section here is pretty straightforward. See the arithmetic reasoning portion of the guide for examples of perimeter and area. New information will be covered here.

Angles

ANGLES are typically measured in degrees. As an example, if you were to completely rotate around a circle a single time, you would have gone 360 degrees, since there are 360 degrees in a circle. Half a circle is 180 degrees, a quarter circle is 90 degrees, and so on. Degrees are used to talk about what fraction of a total rotation around a circle a certain angle represents.

Table 8.2. Types of angles

TYPE OF ANGLE	DEFINITION
Acute angles	Angles with a measurement of fewer than 90 degrees
Right angles	Angles with a measurement of exactly 90 degrees
Obtuse angles	Angles with a measurement of more than 90 degrees
Complementary angles	Two angles which add up to 90 degrees
Supplementary angles	Two angles which add up to 180 degrees

Note that the sum of two angles that form a straight edge is 180 degrees. Note also that angles corresponding angles of parallel lines are equal.

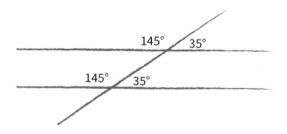

Figure 8.1. Corresponding angles of parallel lines

Triangles

TRIANGLES are geometric figures that have three sides (which are straight). The most important thing you need to know about triangles is that the sum of the measurements of the angles always equals 180. This is important because it means that if you know two of them, you can very easily calculate the third simply by subtracting the first two from 180. There are three types of triangles that you need to know about.

Table 8.3. Types of triangles

Type of Triangle	Definition
Equilateral triangles	These are triangles in which the three angles are the same measure. Each one of the angles is 60 degrees.
Isosceles triangles	These are triangles which have two sides of the same length. The two angles which are directly opposite the sides of the same length will be the same angle. If B was the third angle, then you could state that lines between A and B (AB) and between lines B and C (BC) are the same. AB = BC.
Right triangles	These might be the most important. A right triangle is a triangle which has one side that equals 90 degrees. Two sides are legs and the third side, which is directly opposite the 90-degree angle, is the hypotenuse, the longest one of the sides.

With right triangles, you can always remember that the lengths of the sides are related by the following equation (the Pythagorean Theorem, for those of you who remember):

$$a^2 + b^2 = c^2$$

Remember that c is the hypotenuse, which is the side opposite the right angle. It is the longest of the sides.

The area of triangles is $\frac{bh}{2}$, where the base is perpendicular (form a right angle) to the base.

Also, with right triangle with two legs that are equal, the hypotenuse is always $\sqrt{2}$ times the value of a smaller side. For right triangles with 30-60-90 angles, the hypotenuse is always twice the smaller side (side across from the 30 degree angle), and the other leg (across from the 60 degree angle) is $\sqrt{3}$ times the smaller side.

Circles

A CIRCLE, as you probably are already aware, is a closed curve where every single point is the exact same distance from its center. If you draw a line from the outside of the circle to the fixed point in the center, you have the radius. If you draw a line from one side of the circle straight through to the other, passing through the center point, you have the diameter. A chord is a line from one point of the circle to another, but not necessarily a diameter. By the definition of a circle, all radii (plural of radius) are equal and all diameters are equal.

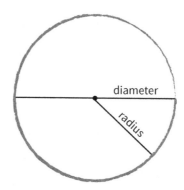

Figure 8.2. Diameter and radius

For a circle, remember that diameter = 2 × radius.

You can use the radius to find out the circumference of a circle (how big it is around), by using the following formula, where C = circumference and r = radius:

$$C = 2\pi r$$

Think of the circumference like the perimeter of the circle; if you were to take a piece of string all the way around a circle and measure it, this would be the circumference.

To find the area of a circle, use the following formula, where A = area and r = radius.

$$A = \pi r^2$$

The area is the measurement of what's inside the circle and is written in square units.

The Coordinate System

The coordinate system or, sometimes, the Cartesian coordinate system, is a method of locating and describing points on a two-dimensional plane; It is a reference system. The plane is two number lines that have been laid out perpendicular to each other, with the point that they cross being origin (0,0). The origin is the point 0 for both the x (horizontal) and y (vertical) axes. Positive and negative integers are both represented in this system.

In the Figure 8.3, each small tick on the line is equal to 1. The larger ticks represent multiples of 5. A point is also depicted, P, which shows how things are placed onto the coordinate system. Again, the horizontal line on the coordinate plane is called the x-axis, and the vertical line on the coordinate plane called the y-axis. The points, when described, are described in reference to where they lie on that plane.

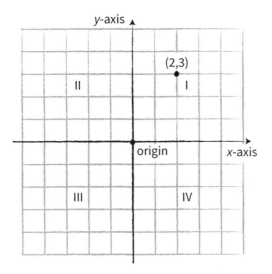

Figure 8.3. The coordinate system

Here is an example that is the point shown above:

$$(2,3)$$

The first number, 2, is the x-axis. The second, 3, is the y-axis. So this point is at position 2 of the x-axis (go over 2 units to the right) and position 3 of the y-axis (go up 3 units). Note that for negative x values, we go to the left that many units, and for negative y values, we go down.

A way to simplify the way these points work is to say:

$$(x,y)$$

Slope

The SLOPE is the steepness of a given line. When you look on the coordinate plane, if you draw a line between two points, you will get the slope. One of the main uses of a slope is to signify a rate; it measures how much a certain thing goes up for every unit it goes over.

Here is the way to find slope:

$$\text{Point A: } (x_1, y_1)$$
$$\text{Point B: } (x_2, y_2)$$
$$\text{Slope} = \frac{y_2 - y_1}{x_2 - x_1}$$

In plain English, you will take the difference between the y-coordinates and divide them by the difference between the x-coordinates. Remember: y is vertical and x is horizontal.

The following is an example of how this type of problem would likely manifest:

Some things to remember:

- If the slope is positive, the line is going up as it goes right.

- If the slope is negative, the line is going down as it goes right.

- The larger the absolute value (positive value) of the slope, the steeper the line.

- If a line is horizontal, its slope is 0; if a line is vertical, its slope is undefined.

- The formula for a line is $y = mx + b$, where m is the slope and b is the y-intercept (where the line crosses the y-axis).

- If two lines are parallel, they have the same slope. If they are perpendicular, they have negative reciprocal slopes.

- If two lines are intersection, the x and y values work in both equations; in other words, if you were to plug in the x and y values, both equations would be true. This is how you "solve" a system of equations.

Tips

Here are some tips to help you make it through the mathematics knowledge section of the ASVAB:

- Remember to utilize PEMDAS and the order of operations when you are working through problems.

- The more you practice, the easier the problems will be. There are a lot of internet sites with math problems.

- Be careful with the answers. The test will often provide common mistakes answers among the correct answer.

- Since you will not have access to a calculator, you will need to round π, 3.14, or $\frac{22}{7}$ are the most common ways to do this. Sometimes you may be asked to simply include it in your answer without actually utilizing the digits in your calculations at all.

- Don't mix up the perimeter and area formulas.
- One way to get good at algebra relatively quickly is to set up all of your problems, even simple ones, as algebra problems. Remember, you can simply create a variable and stick it on one side of the equation to solve for it. For example, *3 times a number added to 10 is 16* becomes $3x + 10 = 16$. Translate almost word-for-word from English to math, and then get x by subtracting 10 from each side, and then diving by 3.
- Keep in mind that, unlike many tests you probably remember taking in high school, you are not being tested on your ability to write out how you solved the problem. There are many correct ways to solve mathematical problems and, as long as you come to the right answer in the end, it does not matter how you solved it. Nobody will be coming behind you and checking your scratch paper to see what you did (unless they think you are cheating, but that is a whole other ball of wax).
- If you can't solve the problem directly, trying plugging in the answers to see if you can figure out which one works.
- Pay close attention to the positive and negative signs in the work that you are doing. This is extremely important because, again, they will likely throw the correct answers out there as one of the choices but with the wrong sign.
- Always make sure the fractions that you are working with are simplified when you finish with them. Unless otherwise stated, the test answers will usually want the most simplified version of the fraction.
- Make sure you go step by step through each question. Don't skip steps or combine steps. Doing either could lead to an issue where something is accidentally missed.
- It might help you to change even the normal expressions into algebra. Often, understanding what you are being asked to do and knowing how to handle certain problems is only made easier when you are using algebra to handle it.
- Draw pictures on geometry problems.

GO ON

Practice Questions

1. Evaluate the expression $\frac{4x}{x-1}$ when $x = 5$.

   A.   3

   B.   4

   C.   5

   D.   6

2. Simplify: $3x^3 + 4x - (2x + 5y) + y$

   A. $3x^3 + 2x + y$

   B. $11x - 4y$

   C. $3x^3 + 2x - 4y$

   D. $29x - 4y$

3. Simplify: $(x + 7)(x - 5)$

   A. $x^2 + 2x - 35$

   B. $x^2 \pm 2x - 35$

   C. $35x$

   D. $7x^2 + 5x - 35$

4. Simplify the expression $\frac{4xy^3}{x^5y}$.

   A. $\frac{12}{x^4}$

   B. $12(x^2y)^2$

   C. $64(x^2y)^2$

   D. $\frac{64y^2}{x^2}$

5. Find the area of a rectangular athletic field that is 100 meters long and 45 meters wide.

   A.   290 m

   B.   4,500 m²

   C.   145 m²

   D.   4.5 km²

6. Mary runs 3 miles north, 4 miles east, 5 miles south, and 2 miles west. What are her final coordinates (in miles), with respect to her starting point?

   A. $(8, 6)$

   B. $(-2, 6)$

   C. $(7, 3)$

   D. $(2, -2)$

7. Two identical circles are drawn next to each other with their sides just touching; both circles are enclosed in a rectangle whose sides are tangent to the circles. If each circle's radius is 2 cm, find the area of the rectangle.

   A.   24 cm²

   B.   8 cm²

   C.   32 cm²

   D.   16 cm²

8. Solve for a: $3a + 4 = 2a$

   A. $a = -4$

   B. $a = 4$

   C. $a = \frac{-4}{5}$

   D. $a = \frac{4}{5}$

9. Evaluate the expression $|3x - y| + |2y - x|$ if $x = -4$ and $y = -1$.

   A. -11

   B.   11

   C.   13

   D. -13

10. Solve for x: $8x - 6 = 3x + 24$

    A. $x = 3.6$

    B. $x = 5$

    C. $x = 6$

    D. $x = 2.5$

11. Convert 0.25 into a percentage.

    A.   250%

    B.   2.5%

    C.   25%

    D.   0.25%

12. $-2x - 3x^2 + 4x =$

    A. $-3x^2 - 2x$

    B. $-3x^2 + 2x$

    C. $3x^2 + 2x$

    D. $3x^2 - 2x$

13. Convert $\frac{38}{98}$ into a percentage. Round to the nearest percent.

    A.   39%

    B.   38%

    C.   38.77%

    D.   38.775%

14. Factor the following: $x^2 + 4x + 4$

    A. $2(x + 2)(x + 2)$

    B. $(x - 2)(x - 2)$

    C. $(x - 2)(x + 2)$

    D. $(x + 2)(x + 2)$

15. Solve for x: $4x + 3 > -9$

    A. $x < 3$

    B. $x > -3$

    C. $x > -1\frac{1}{2}$

    D. $x < 1\frac{1}{2}$

GO ON

Math Knowledge Answer Key

1.	C.	9.	C.
2.	C.	10.	C.
3.	A.	11.	C.
4.	D.	12.	B.
5.	B.	13.	A.
6.	D.	14.	D.
7.	C.	15.	B.
8.	A.		

Review and Takeaways

The mathematics knowledge section is meant to test your ability to complete basic high school level mathematical operations.

Review

- **NUMBER THEORY:** All about numbers: basics of numbers and number systems, the order of operations, different operations, and numerical operations.
- **FRACTIONS:** How to handle fractions: operations and manipulation of fractions, including how to convert them to percentages and decimals.
- **ALGEBRA:** An overview of basic algebra, using variables: how to use variables in word problems, and how to solve for these them.
- **GEOMETRY:** Shapes: circles, squares, triangles, and angles: How to get perimeters and areas. This also includes slope and the coordinate plane.

Takeaways

The most important thing to take away from this section is that practice makes perfect. If you do a quick online search, you will find thousands of examples and practice problems to help you get through this section. Start working on them in your free time. You can also just make up problems and equations when you get a free minute and then solve them on your own.

nine

ELECTRONICS INFORMATION

Introduction

The electronics information subtest is one of the technical tests. It is not a required test to be admitted into any branch, though it does help to determine which jobs you will be eligible for once you have joined. If your career is going to be focused on the manipulation or use of electronics (or if it simply requires a high score on this subtest) then it would behoove you to spend a little bit of extra time on this section and learn it thoroughly.

This subtest concerns itself primarily with the study of electricity and its uses in modern technology. Along with the information about electronics itself, you will also need to have a good understanding of mathematics and algebra. Both of those disciplines often coincide with the information you learn about electricity and electronics (particularly algebra, which forms the basis for literally all of the equations utilized in the study of electronics).

Electricity: The Basics

One of the most striking things about modern society is the intense reliance on electricity and technological innovations which use electricity to function. Electricity is the power from which so many household, industrial, and technological items derive their ability to function. Electric charges come in two basic types, each other their own purposes and their own distinct characteristics. **POSITIVE ELECTRIC CHARGES** are in the form of protons, which are positively charged subatomic particles. **NEGATIVE ELECTRIC CHARGES** are in the form of electrons, which are negatively charged subatomic particles.

Electric charges exert a force on other charges in their vicinity through the creation of an electric field around themselves. Charges which are the same (positive and positive, negative and negative) repel one another. Charges which are opposite of each other (positive and negative) attract each other.

Charges which are moving are able to exert a force on other moving charges in the form of a magnetic field. Magnetic fields that change over time are able to create electric fields through a process that is known as electromagnetic induction. Likewise, electric fields which vary over time are able to create magnetic fields. Both of these types of fields are different forms of energy. Charges that are being accelerated create changing fields of both types which travel farther than the subatomic particles which originally resulted in their creation. The term "electro-magnetic" is often used to describe the relationship and fluctuations between changing electric and magnetic fields.

Electromagnetic waves are the force fields of disturbance which emanate from the fields created by charges. You know a few of these types of waves already from your day to day life: x-rays used in medical centers, radio waves, light waves, and many other are all types of electromagnetic waves. These can travel through different types of matter depending on the frequency of the waves. Light, for instance, can travel in air or water but will not pass through solid objects. X-rays will pass through solid objects, which is why they are useful for examining the interior of compartments (or the human body). Radio waves can pass through some solid objects but can't pass through water. Electromagnetic waves do not need any medium in order to be created or to move. Thus they are differentiated from, for instance, sound waves. They travel through a vacuum at the speed of light, which is 186,000 miles per second.

Here are some basic definitions you will need to have in mind moving forward through this section of the guide:

- **AMPERES**: Amperes (amps) are the number of electrons moving past a given point in one second.
- **OHMS**: Ohms are used to measure resistance and are used to describe the way electron flow is being limited.
- **VOLTS**: Volts are units which are used to measure how much work is being done when electrons are moving from one point to another point.
- **CURRENT**: Current is a word used to describe electron flow (electrons moving from one point to another point).
- **CONDUCTORS**: A conductor is anything which works to assist electric current in its ability to flow.
- **INSULATORS**: An insulator is anything which works to prevent electric current from flowing.

- **WATTS**: Watts are units which are used to measure how much energy is being consumed. Related to this are watt-hours, which measure the amount of energy used in an hour. The most common way to measure electricity is to measure it in kilowatt hours, the amount of energy which is used in a single hour by a single kilowatt of electricity.
- **CIRCUIT**: A circuit is a series of components through which electric current flows.
- **IMPEDANCE**: An impedance is anything (including components) which changes the way the current flows.
- **RESISTANCE**: Resistance is a type of impedance. In resistance, electrical energy is converted into heat.

When people talk about electricity, they are almost always talking about electrical currents. Electric current is a term which is used to describe electrical flow. The direction of the current is defined as the direction in which the positive charge is flowing. Conductors, which assist with the flow of electrons, are usually some form of metal. In that case, the charges which are being moved are electrons, negatively charged subatomic particles. Since only electrons are able to move in metals, the direction of the current is the opposite of their flow.

Not all materials are created equal when it comes to their conducting (or insulating) characteristics. Wood and glass are not good for conducting and, in fact, work as insulators. Metals, particularly copper, work wonderfully as conductors, allowing the free flow of electrons. There are some substances, known as semiconductors, which do not have free electrons but with which electrons can easily become free in order to allow for current. Silicon is often used to make some semiconductors as well as for the creation of electronic components like transistors.

For electric current to flow, it must have somewhere to flow to. You can think of this as a circle which needs to be completed for any flow to occur. Flow is continuous or it is not happening at all. Charges, thus, do not build up in a single place except during special situations. The simplest circuit will consist of a power source and a component which is using power. Current will flow from the power source to the component (known as the load) and then back to the power source. This is known as a complete circuit.

An open circuit, on the other hand, is one in which electrons do not flow back to the power source. A short circuit is one in which something closes the circuit so that the current is unable to get to the load (it is stopped short). Short circuits are unintended and usually cause issues with the circuit and can lead to power failures and other hazardous situations.

Electrons do not flow through a circuit of their own accord. They have be moved by way of some force. The force that is used to move those electrons is the voltage, which is provided by the power source that is being used.

Power

POWER is generally defined as the rate that work is being done. Another way to say this would be to say that power is the amount of energy being used per unit of time. In the case of electricity, power is used to state the rate that electric energy is being moved by a circuit. In the above definitions, you see that the watt is the unit used to describe power. One watt is one joule per second.

ELECTRICITY is generated in a number of ways. It can be created through the burning of fuels, as is the case with the burning of fossil fuels (coal, oil, etc.) or nuclear. The way that the power is generated by these devices (known as generators) on a large scale is relatively simple: Water is held inside of a boiler. The water is then heated so that it will turn a turbine. That turbine, in turn, begins turning the generator itself, which is what produces electric power. This process usually results in the creation of environmentally damaging products, including hydrocarbons and other greenhouse gasses.

To offset that potential, there are alternative energy sources which have been created, including wind turbines and hydroelectric energy sources. The benefit of these types of generators is that they do not require any fuels to be burned in order to turn their turbines and generate electric power. The negative side, however, is that they often rely on somewhat unpredictable natural phenomena. Additionally, they have to be put in very specific locations to work properly. Wind turbines use the power of the wind to turn the turbine by catching it on fins. Hydroelectric sources use natural water flows going through a turbine in order to turn a generator and produce the electricity.

There are two types of electricity which are generated, both with advantages and disadvantages. DIRECT CURRENT is a one directional flow of current. This is the type of power being products by solar cells, batteries, and some other power generating devices. ALTERNATING CURRENT is a flow of current that reverses its direction of flow on a periodic basis.

When voltage is created and output by the generator, it is usually increased orders of magnitude higher than it originally was so that it can be transmitted through the power grid, where households, buildings, and other facilities can accept the power and use it for their own devices. Typically, when the high voltages need to reach homes and other buildings, the voltage is lowered again at local substations. Homes and other buildings usually have "transformers" on the electrical poles outside. These are used to further lower the voltage so that it can be utilized safely in homes.

Components, Symbols, and Units

When discussing electricity and electronics, it is important to have an understanding of the basic units which are used. The following table should clear a bit of that up:

Table 9.1. Basic electronic units

QUANTITY	UNIT	SYMBOL
energy	joule	J
power	watt	W
charge	coulomb	C
voltage	volt	V
current	ampere	A
frequency	hertz	Hz
capacitance	farad	F
inductance	henry	H
resistance	ohm	Ω

Additionally, there are a number of common components which are used across the board in the study of electronics. Here is a brief list of them, along with their descriptions:

- A SWITCH: An electronic component which can be used to either divert current from one conductor to another conductor or to stop the flow of the current.

- RESISTOR: A two-terminal electronic component that is used to add resistance to an electrical circuit.

- INDUCTOR: A two-terminal electronic component which resists any changes in the current which passes through it.

- DIODE: A component which is used to produce asymmetric conductance. It allows current to flow in only one direction.

- FUSE: A fuse is a component which is rated for a specific current and is used to prevent the flow of electricity/current from hitting a level above that rating.

- GROUND: A ground is the most common return path for an electrical current. This is the point from which voltages are measured. This also helps prevent unwanted transference of electricity.

- BATTERY: A battery is, usually, an electrochemical cell which is used to store energy in the form of chemicals and then, when in use, to change that chemical energy into electrical energy.

- CAPACITOR: A capacitor is used to store energy through electrostatics.

- **TRANSISTOR**: A transistor is used to amplify and modify electronic signals and current.
- **SOLDER**: Solder is a fusible metal allow with a low melting point (one lower than that of the pieces that it is being used on) which is used to join together components.
- **FLUX**: Flux is a substance which is used to help increase the both the melting and the flow of metal. It is used along with solder to help facilitate metal joining.

When used in circuit and electrical diagrams, these components all have a unique appearance and a short hand notation. Below are all the symbols used:

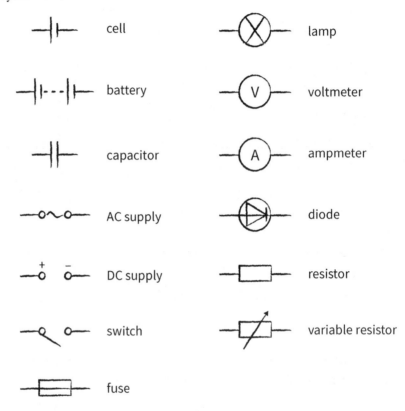

Figure 9.2. Circuit and electrical diagrams

See Figure 9.3 for examples of some of these components, showing their relative sizes:

Overall, these parts are self-explanatory. There is, however, need to describe how solder and flux work in the system as a whole. Solder is used to connect components together. Its low melting point allows it to be used with relative ease and safety without causing damage to the other parts you are working on. Usually, solder is used to connect new wires to a circuit or to hold down components onto a circuit board. Flux is used to help clean oxidation off of the wires which are being connected.

Solder is also used for some other purposes and has a different makeup. Acid flux is put into some types of solder used for industrial applications. This type of solder is not utilized in electrical circuits because it has the potential to corrode the wires and cause a short circuit.

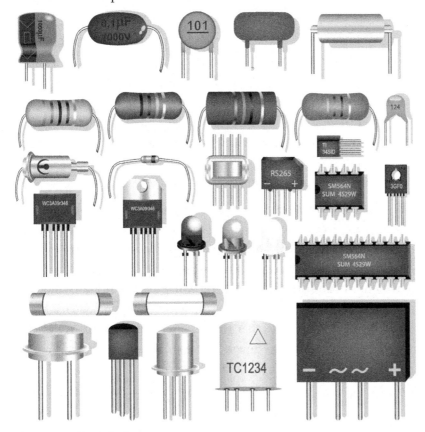

Figure 9.3. Electrical components

Another thing you will want to do when attempting to understand electricity is to review the way that very small and very large quantities are written through the metric system and with scientific notation (discussed in the general science subtest portion of the guide).

Alternating Current vs. Direct Current

Current comes in two forms: oscillating (alternating current) and constant (direct current). The most common sources of DC power are batteries and solar cells. Power outlets are the sources of AC power. This is because the power generated in most power plants is transmitted in the form of alternating current. The oscillation of AC current has a voltage which repeats multiple times per second, usually at a frequency of either 60 Hz or 50 Hz. In North America (particularly Canada and the United States), AC outlets have a voltage of 115 V and they have a frequency of 60 Hz. In European countries, voltage is between 220 V and 250 V at 50 Hz. This is why the plugs used in outlets differs between European countries and the United States.

Modern electrical outlets have three slots for prongs to enter. The smaller flat slot is for the positive flow (the "hot" terminal). The longer flat slot is the neutral terminal. The circular, third slot, is for negative flow and is a grounding terminal. Likewise, when you look at the prongs on a plug, you will see a narrow, wide, and third (U-shaped) prong. These are equivalent parts to the terminals on the outlets themselves.

There are two phases through which power is delivered:

1. Single phase: Indicates an oscillating voltage between the hot and the neutral terminals.
2. Three phase: This is used for higher voltages and indicates the presence of three terminals which are "hot". All three of these terminals have the same oscillating voltage. The current flows between each phase and the two terminals. This is often the type of power utilized in industrial machines and settings.

AC has one large advantage over DC: The amount of voltage being produced can be modified very easily through the implementation of a transformer. Transformers are made through the use of a primary and a secondary inductor which are wrapped around a core made of iron. The side which is nearest the source of the voltage is the primary inductor. The other side is the secondary inductor. More turns will equal a higher voltage. If the primary is higher than the secondary in the number of terms, the voltage will step down. If the secondary is higher than the primary in the number terms, the voltage will step up.

The way that most generators make AC power is through the use of coils which are rotating inside of magnetic fields. That rotation produces a voltage which varies with time. DC power is created by using a rectifier, which converts AC to DC.

Laws: Ohm and Joule

There are two figures which tower miles above many others when it comes to electronics and electricity in general: Ohm and Joule. The contributions that these two men made ultimately led to the immense advances in the field of electronics that you see today.

Ohm described the way that voltage, current, and resistance relate to each other. His law, **Ohm's law**, described that the resistance multiplied by the current will yield the voltage. The following equation describes this principle, where I = current, V = voltage, and R = resistance:

$$V = IR$$

It is important to keep in mind how circuits work and how this law relates to them. In a normal circuit, the current will be flowing from a positive energy source to a negative energy source. Since there is only one path for the current to go, it is the same all throughout the circuit. With the inclusion of a resistor, the voltage will be modified.

For an example of how this might work, take a look at the following sample problem:

Though this equation looks fairly simple, it is also robust enough to be used in more complex circuits. Components of circuits are connected in one of two ways. In a **PARALLEL** circuit, current is divided between components, while current flows equally through all of the components in a **SERIES** circuit.

Obviously, when this includes resistors, it changes the way that the resistance is calculated. Fortunately, there is a simple way for you to determine the resistance even when there is more than one resistor so that you can utilize it in the equation above.

For resistors that are connected in parallel:

$$\frac{1}{R} = \frac{1}{R_1} + \frac{1}{R_2} + ...$$

For resistors that are connected in series:

$$R = R_1 + R_2 + ...$$

The examples here indicate (via ...) that you can do this for however many resistors are being used in the circuit.

The next towering figure in electronics is Joule. **JOULES LAW** provides a way to determine how much power is being lost when the current goes through the resistor. The power which is lost is the product of the voltage going across the resistor and the current which is going through the resistor.

Here is the equation for Joule's law, where P = power, I = current, and V = voltage (resistor):

$$P = IV$$

Additionally, there is a way to change out the voltage which is used in Joule's law for the equation in Ohms law. For the purposes of that, remember Ohm's law: $V = IR$.

Next, add in Ohm's law to Joule's law in place of V.

$$P = I \times I \times R$$

Leaving you with the following equation:

$$P = I^2R$$

This equation is useful when you do not have the voltage but you do have the resistance and current.

Most appliances used for heating which are found in the home utilize what is known as resistive heating, which is the use of a wire with high resistance to allow excess electricity to leave the circuit as heat. This is a resistor which is used in these is known as a filament. This is the same way that incandescent light bulbs work (and the reason why they are so hot when you touch them after they have been on for a while).

Electric Devices

Motors

As you probably already know, electric motors are used in a variety of modern devices. Automobiles have a number of motors to fulfill different functions. Vacuum cleaners, dryers, and other consumer electronics also have motors (of differing sizes). A motor, at its most simple, is a number of coils of wire around a magnet. When current gets pushed into the coil it creates a magnetic field which then causes the motor itself to turn. The current in the coil will reverse each time the coil passes by the magnet. This, in turn, causes the force to reverse direction and repel the coil.

This principle works two ways. With **AC CURRENT**, the current will already be oscillating direction so there is no need to modify the force's direction. With **DC CURRENT**, something has to reverse the current direction twice every time it revolves around the motor.

AC motors come in two basic types. A **SYNCHRONOUS** motor is one which automatically goes with the oscillations of the current coming into it. It turns at the same rate as the frequency of the alternating current. An **INDUCTION** motor is one where the coils are surrounding a conductor instead of a magnet. The electromagnetic induction then makes the current flow through the conductor because of variations in the magnetic field that is being created by the coils.

Connectors

On printed circuit boards, components are often linked in circuits. This is the way that most electronic devices and computers work. The boards are comprised of an insulating material, usually some sort of resin or hard plastic, on which a conducting coating has been placed. The coat is usually copper. Once the copper has been placed on the board, acid is used to etch away the excess copper. What you are left with after the acid etching is the final layout of the conducting strips which connects the components to each other on the circuit board. Traces are the most common name associated with those strips.

Figure 9.3. Traces on a printed circuit board

Wires are a common link between components of circuits. They conduct electricity and they are used primarily because they are easy to manipulate. In essence, a wire is ideally going to be the minimal piece of metal that is required to make a useful connection between two electronic components. **SOLID** wires are single pieces of metal which are used to connect components. **STRANDED** wires are formed of multiple thin solid wires which have been twisted together in order to make one conductor.

Of the two, stranded are the more common. They are both more resistance to breakage than solid wires and they have additional flexibility. Conductors can be insulated as well. This allows them to move current without losing any to the environment that they are running through. It also drastically reduces the potential for a short circuit. You have probably seen this in action: the insulator would be the little sheath of plastic which is around the wire itself. Insulated wires are almost universally used over "bare" wires (wires without any insulation).

Not all connections will be able to work with the use of a simple wire. In those cases, the need for a more complex conduction mechanism comes into play. Single conductors, for example, have issues with "noise" which can enter from either component in the actual circuit itself or from the environment in which the conductor is found. Signal wires which are found close to their return path can minimize this issue. Twisted pairs, where insulated conduction mechanisms are coiled around each other, make use of this principle. This is the way most Ethernet cables work. If you were to look at the jack on the end, you would see the wires inside twisted around each other to reduce noise.

In order to get even more isolation from environmental noise, coaxial cables are sometimes used. These are the types of cables that you might see coming out of your television from your cable box. They are made of three separate sections: The conducting wire in the center, a dielectric insulator which surrounds the conducting wire, and an outer "shield" which is comprised of braided conducting material. The outer conductor is meant to absorb and redirect any environmental noise that would otherwise affect the signal itself.

Vacuum Tubes

Many of the more complex electronic devices that are produced need some sort of way to control how the electric current is flowing. Now that semiconductors exist, there is an easier way to accomplish this. In the past, however, a component known as a vacuum tube was used for this. Vacuum tubes are tubes which have air removed from them. The shell of these tubes is typically made of glass. Inside of the tube, a cathode (a metal device used to emit electrons) sends out electrons that an anode (another metal device) can accept. This is accomplished by first heating the cathode so that it begins emitting electrons. Afterward, an electric field is created between the two parts which push electrons in the direction of the anode. Electrodes which are inside of the vacuum tube help to direct the way that the electrons flow.

Diodes are a simple type of vacuum tube.

In modern electronic devices, vacuum tubes are no longer in use. Semiconductors have replaced them.

There are, however, some specialized situations in which vacuum tubes are preferred. One of these situations is with guitar amplifiers, which produce a different sound when vacuum tubes are utilized in their construction. That sound is preferred by many enthusiasts.

Semiconductors

Semiconductors are materials that can work as either insulators or conductors depending on how they are used and other external factors. A number of external factors can be applied to a semiconductor to change its conducting properties, including lights, voltage, or other things. Transistors are a good example of an electronic component which utilize semiconductors. Commonly, semiconductors are considered to be any materials which have a conductivity rating that is between an insulator (plastic, glass, etc.) and a conductor (copper, etc.).

Here is a brief list of common semiconductors:

- gallium arsenide
- germanium
- selenium
- silicon
- some alloys
- some oxides
- silicon carbide

The vast majority of semiconductor use takes place in integrated circuits. These types of circuits are the ones which are used in most contemporary electronic devices; cell phones, laptops, computer chips, etc.

Integrated Circuits

Most modern electronics make extensive use of **INTEGRATED CIRCUITS**. These are also known as *computer chips*. They are a method of creating high-density electronic devices for use in other machines. A flat silicon core, cut to the required shape, is coated with some sort of material which will insulate it. A series of paths for conduction and electronic components are then inlaid onto that insulating layer. The silicon acts as a semiconductor.

The design of these chips is very similar to the way that PCBs are made (you will recall PCBs are printed circuit boards, which were covered earlier in this guide). Photolithography is the process through which these integrated circuits are etched. Multiple types of electronic components can be created in this way for use on integrated circuits. Thus, anything that can be made using electronic components can usually be made on a chip as well.

Computers utilize integrated circuits for nearly all of their operations primarily because these circuits can be used to do arithmetic operations which are at the heart of everything a computer does. Integrated circuits that contain many transistors are utilized for these purposes and are known as microprocessors. The way computers store data utilize these. RAM and ROM are terms you have probably heard when dealing with computers. **ROM** (read-only memory) is a static type of memory that stores information regardless of whether it is powered or not. Data stored here is written when chips are created and is read by the processor when it needs to be accessed. **RAM** (Random access memory) is a dynamic memory that is refreshed continuously and, when the chip ceases to be powered, is lost. This is the memory which is modified by the processor.

Media

Analog vs. Digital

If you have spent any time at all around electronics, you have probably heard the terms *analog* and *digital* used to describe the way some devices work. Here is a breakdown of the two. **ANALOG** devices are devices which have an always changing voltage. **DIGITAL** devices are devices in which the voltage is a discrete value.

Sounds complicated? It isn't. Analog devices are fairly simple to understand. Digital devices have a more complex functionality. Signals are usually in analog form. Digital devices will take analog signals (usually) and then interpret them so that they can be stored and modified. Digital devices work through the use of a binary code, which is a fancy term for the 1s and 0s you may have heard about when dealing with computers. When needed for human use or interpretation, digital signals are then converted back to analog so that people can use them. Almost universally, interfaces that humans use are going to be in analog form.

Most devices that work through analog will have a digital version as well. You are familiar with this, as most analog devices are rapidly being replaced by their digital versions. VCRs and VHS tapes are being replaced by DVD players and DVDs (or Blu-Ray discs). Computers operate largely in the digital space. They store information in binary and when the information needs to be presented to the end user it is converted to analog as an image on the monitor.

The complexities of analog and digital devices are much deeper than what is being presented here, but the basics are the only things you will need to be aware of for the purposes of the ASVAB subtest.

TV and Radio

Everyone is familiar with television and radio, two of the most popular methods of communication in the modern world. These utilize electromagnetic waves due to the speed at which they travel. Radio, of course, was developed first.

Using amplitude modulation (AM), radio originally operated much like Morse code does. It was a series of on and off signals which could then be interpreted. By changing the way that the EM wave is traveling to match the signal, complex signals such as music can be transmitted as well. This is a byproduct of the modification of the size of the electric field over time.

The signal being sent is the modulation. The EM wave itself is known as the carrier. When you hear a radio broadcasting an AM signal, you hear the sound at a frequency which is equal to 1 MHz. The carrier frequency is the radio station, the one which the station is using to identify its signal in comparison to others. The frequency of the signal is 5 kHz, which is a small part of the complete audio band.

FM stands for frequency modulation, and is the method by which television and FM radio transmit their signals. FM is going to change the frequency of the EM wave being used to transmit the signal. The amplitude does not change, but the frequency does. Most FM radio is going to modulate the frequency to around 15 kHz. This modulation gives it a higher quality sound than AM radio is able to provide.

This same method of transmission is utilized by some other communication devices as well. Cell phones, wireless routers, and other devices which transmit without wires also modulate the frequency.

Discs

Media, rather than being stored on magnetic devices or paper as was common in the past, is now stored in discs. These include compact discs, digital video discs, and Blu-ray discs. The density that contemporary media requires for storage (and space considerations) has led to the creation of these discs for easier and more convenient storage and distribution. Small pits are created (on the micro scale) on small plastic discs that are (usually) about 10 cm across. Lasers which then shine on these discs from the devices reading them will reflect in a different way from the pits that are on the disc. The reading of those reflections will then create a digital signal which can be used to display the media.

All of the differing types of discs can hold varying amounts of information:

Table 9.2. Common information storing devices

DEVICE	DEFINITION
CDs	Around 700 megabytes of data will fit onto the average CD.
DVDs	DVDs can hold around 4 gigabytes of data.
Blu-ray	These discs have a data capacity approaching 25 gigabytes.

It should also be noted that these discs can be double sided, where each side of the disc contains its own information. This allows for twice the storage space on a single disc.

Common Electronic Tools

A variety of tools are utilized when working with electronics. This is because many of the systems which are being worked on are sensitive and have to be consistently measures for a variety of metrics.

The following are some of the most common electronic testing tools you will encounter:

♦ MULTIMETER: A multimeter is a digital electronic testing tool which can be used to measure a variety of things including voltage, resistance, and current.

♦ VOLTMETER: The analog version of a multimeter.

♦ OSCILLOSCOPE: An oscilloscope is a device which is used to display changes in the voltage over a period of time.

♦ FREQUENCY COUNTER: A frequency counter is used to measure frequency.

♦ OHMMETER: These devices are used to measure resistance.

♦ AMMETER: These devices are used to measure current.

- ◆ **PULSE GENERATOR:** This is used to create rectangular pulses to test circuits.
- ◆ **POWER SUPPLY:** Power supplies are used to supply electrical energy to a load device.
- ◆ **SIGNAL GENERATOR:** A signal generator is a device which create electronic signals either in analog or digital form.
- ◆ **DIGITAL PATTERN GENERATOR:** This is used to make stimuli for digital electronics.

Tips

Don't second guess yourself. If you know what current is, don't let them confuse you with some of the additional multiple choice answers that they toss at you. Often, they will put answers in there which are close to the correct answer but backward. Keep your wits about you.

Make sure you have your algebra down. The equations here may be simple, but that does not mean they will not have to be manipulated at times to yield the results that you need them to. Looking at Ohm's law, for instance, you may have to solve for current or resistance rather than the voltage which is being solved for in the standard version of the equation.

One of the best things that you can do to learn this information quickly is to play around with electronics on your own. Multimeters, for instance, do not cost a great deal of money. You can pick one up for relatively cheap and then use it to test out some of the circuits around your house. It is perfectly safe and it will pay off in the long run.

Building circuits is something which is extremely easy (and fun) which will also teach you plenty about the process of electronics as well.

Practice Questions

1. Electricity forms the basis for much of the technology which is utilized in the modern world. Which of the following can be used to produce electricity?

   A. wind

   B. atoms (nuclear material)

   C. coal

   D. all of the above

2. What is the purpose of a capacitor in a circuit?

   A. increase voltage

   B. get rid of excess energy (dissipate charge)

   C. store electrical energy (accumulate charge)

   D. speed up the current capacity

3. What needs to happen to the voltage produced by a power plant before it reaches a residential home or an office?

   A. It needs to be stepped up.

   B. It needs to be stepped down.

   C. It needs to be dissipated.

   D. It needs to be used.

4. There are two types of current which are product: AC and DC. What is the difference between the two?

   A. AC current is constant and DC current oscillates.

   B. AC current is only made by batteries.

   C. AC current is only made by solar cells.

   D. AC oscillates its current and DC current is constant.

5. What is the purpose of a fuse?

   A. make sure the current does not exceed a specific value

   B. increase the current

   C. decrease the current

   D. allow unidirectional flow of current

6. SI units are the standard for scientific measurement. What is the SI unit that is used to measure resistance?

   A. Ohm

   B. watt

   C. volt

   D. current

7. Solder is often used to connect multiple electrical components together. What is solder composed of?

   A. an alloy of tin and lead

   B. copper

   C. allow of gold and lead

   D. allow of steel and mercury

8. Which of the following methods of producing energy do not produce hydrocarbons which are bad for the environment?

   A. fire

   B. coal

   C. wind

   D. oil

9. Volts are a unit used to describe voltage. Voltage measures how much _____ is being done per unit of charge when an electron moves between two points. Fill in the blank:

   A. energy

   B. work

   C. potential

   D. current

10. All devices utilizing wires receive something after they have their functional tests finished. What is it?

    A. solder

    B. current

    C. voltage rating

    D. codes

11. _____ is a material who does not have a free flowing internal electric charge and which does not conduct electricity.

   A.   conductor

   B.   metal

   C.   plug

   D.   insulator

12. What is the SI unit that is used to measure frequency?

   A.   waves

   B.   electrons

   C.   hertz

   D.   ohms

13. What happens when two wires cross each other and the electricity is able to bypass the circuit it was intended to follow?

   A.   electric leak

   B.   fire

   C.   electrification

   D.   short circuit

14. A common electronic tool is used to compare voltages and then switch the output to show which voltage is higher. What is it called?

   A.   comparator

   B.   voltmeter

   C.   currentometer

   D.   ammeter

15. In the United States, the electricity that comes from power outlets comes in the form of what?

   A.   alternating current

   B.   direct current

   C.   CC power

   D.   heat energy

GO ON

Electronics Information Answer Key

1.	D.	9.	B.
2.	C.	10.	C.
3.	B.	11.	D.
4.	D.	12.	C.
5.	A.	13.	D.
6.	A.	14.	A.
7.	A.	15.	A.
8.	C.		

Review

Electronics and the use of electricity in everyday life have become as ubiquitous as eating and breathing. Understanding the way that these concepts work, particularly with an eye for military service, is important. Though the electronics subtest is not one of the core subtests which will determine whether you can join, it is very important that you are able to understand these concepts for many of the specializations that the military has to offer. Beyond that, it will improve your ability to navigate in a complex and daunting world.

ELECTRONICS BASICS covers the absolute basics of electronics, including information about electricity, some essential background information, and a few equations used in the study of electronics.

The POWER section goes over power or, the rate that work is being done. This is an amount of energy per unit of time.

The SYMBOLS AND UNITS section offeres an overview of the symbols and units which are used in the study of electronics.

AC AND DC are the two forms of electrical current.

The section over OHMS LAW AND JOULE'S LAW offers important equations which relate current, resistance, and voltage.

A MOTOR is a device which is used to translate electrical energy into mechanical energy, usually for motion.

The CONNECTORS section provides an overview of common types of electrical connectors.

The VACUUM TUBES section provides a complete review of vacuum tubes, what they are used for, how they work, and common devices that they are found in.

SEMICONDUCTORS are metals which can either conduct electricity or work as an insulator depending on how they are being controlled by an external force.

ANALOG AND DIGITAL are two different types of signals, their characteristics, pros, and cons.

INTEGRATED CIRCUITS are very small and dense circuits which utilize semiconductors in order to conduct electricity. These are often used in computers and other consumer electronics.

TV AND RADIO covers the method by which AM radio, FM radio, and television signals are transmitted and modified.

Discs are methods of storage. Common types of discs include CD, DVD, and Blu-ray. Small pits are made on the plastic surfaces of the discs which reflect a laser back onto a reading device.

Testing Equipment covers the testing equipment which is commonly used to test circuits and other electronic devices.

Takeaways

What you should leave this section of the guide with is a broad understanding of the most important components, devices, and concepts involved in electronics. Though this test is not one of the core tests which will determine your AFQT score, it is important enough to study if you are looking to get into a specialty job which has a focus on the use of electronics or basic communications tools.

AUTOMOTIVE AND SHOP

Introduction

The automotive and shop information subtest on the ASVAB is used to help calculate some composite scores for specialization and job qualification. You can expect about half of the questions to be about automotive and engine technologies and the other half to be about shop information and tools. The information contained here is relatively basic and does not get too in depth. You will be expected to know how things work in an engine and the very basics about when to use certain tools.

Measurements

To understand engines, you must first understand a few definitions and measurements:

- **BORE:** Cylinder's diameter. In engines, it is used to describe the diameter of the cylinders used to house the pistons.
- **STROKE:** This is a measure of how far the piston can move inside the cylinder.
- **CLEARANCE VOLUME:** This is the volume which is in a cylinder when the piston is in the top dead center position.
- **TOP DEAD CENTER:** The highest position a piston can be in.
- **BOTTOM DEAD CENTER:** The lowest position a piston can be in.
- **SQUARE:** When the bore and the stroke are equal in length. Over or under square describe changes in this (under having a larger bore).
- **ENGINE DISPLACEMENT:** This is the total of all of the piston displacements for each piston in the engine.
- **PISTON DISPLACEMENT:** This is the volume which is

covered by the piston as it moves from the top to the bottom of the cylinder.

◆ **TORQUE:** Torque is a force which twists or turns. The higher the torque the more work can be done.

◆ **WORK:** Work is the movement of some object against a force. Gravity, resistance, friction, and many other forces can play this role. Work is a measurement of force over a distance.

◆ **POWER:** Power is used to describe the rate at which work is being done. Power = torque × speed. Power, in terms of an engine, is described using horsepower.

Engines

The engines which are being tested in the automotive section of the ASVAB are, overwhelmingly, gasoline engines. These engines are known as combustion engines. They combine an accelerant, gasoline, with air inside of cylinders in the engine block itself. The gasoline and air mixture is then ignited. That ignition causes an explosion which pushes the cylinder, turns the engine, and creates the power which is used to make the vehicle move.

Figure 10.1. Engine block

The engine block itself is an extremely sturdy (and heavy) piece of solid metal which is cast using either aluminum or iron. When metal is "cast", it is being melted down and then poured into a mold so that it can take the desired shape (the shape of the mold). Every part of the engine both depends on and stems from the engine block. When you look at an engine, you can see that all of the various parts are attached to the engine block, one way or another.

The head of each cylinder (known as the cylinder head) has a number of features on it, including some small holes which allow any unburned fuel and air into the combustion chamber for use and allows the products of the combustion reaction to escape. These heads are typically either aluminum or iron, with aluminum being the most popular of the two because of its physical properties. Whatever the metal used, it needs to be able to withstand the pressure and heat that is produced by the reaction.

Gaskets are components which are used to help make sure the many parts of the engine fit together. Without these, gasses, oil, or water could leak out of the system. Gaskets come in many shapes and sizes, but they are used to help make sure everything fits tightly together.

There are two components in combustion engines which handle the fuel and gasses. EXHAUST MANIFOLD allows the products of the combustion reaction to leave the engine. INTAKE MANIFOLD allows the fuel and air mixture to get into the cylinder head for use in combustion.

Just below the engine block (but still attached) is the oil pan. This holds oil which is necessary to keep the moving parts of the engine lubricated. This also helps with the dissipation of heat. Without it, the engine would wear itself out very quickly. The bottom of the engine block also has the crankshaft.

The cylinders contain aluminum rods known as pistons. They work much like a gun works. Combustion in the barrel of a gun will create pressure which propels a bullet outward. Like that, the combustion pressure created in an engine pushes the piston down in the cylinder. The crankshaft will then push it back up so that it can be pushed down again in a process which is known as reciprocating motion.

Pistons have to fit very tightly inside the cylinders in order to work properly. This is also how the gasses are kept where they are supposed to be: inside the parts of the engine. Piston rings are the component of the piston which allows them to have the movement that they need while also maintaining a tight seal inside of the engine.

The movement of the pistons is in an up and down motion, correct? So how does that make the tires move? They move in a rotational way, not a reciprocating way. Well, in the same way that the pedals on a bike turn a chain to change the type of motion and make a bike move, pistons in the car engine will turn the crankshaft in order to make the tires move. The pistons are connected to rods which are, in turn, connected to the crankshaft. The primary source of power output in an engine is the crankshaft.

Bearings are placed between the connecting rods and the crankshaft itself in order to allow for the quick rotation and friction caused by the motion of the pistons. The bearings are usually created with a metal that is softer than most other parts used in the engine and they must be kept lubricated with oil.

Even when the engine is idle the crankshaft will be turning. The RPM you see on the gauges inside of your vehicle is a way of describing this turning. RPM stands for revolutions per minute. The combustion chamber will have an explosion every other time the crankshaft rotates. In one minute, this can result in tens of thousands of explosions in the engine each minute it is in operation. The harder the engine is working the more explosions there will be and, thus, the higher the RPM will be.

Even spacing is required for the explosions which are happening inside the chamber because the output of power has to be consistent. A flywheel which is attached to the crankshaft is one of the ways that this is accomplished. The heavy weight of the flywheel will resist changes in its motion and its speed. The size of the flywheel is inversely proportional to the number of cylinders.

Engines vary in the number of cylinders they have and, in fact, this is how they are usually described (based on the number of cylinders and the way that they are being arranged in the engine). Vehicles have four, six, eight, or sometimes twelve cylinders (some vehicles have even more or less—two-stroke engines have two cylinders). You may have heard of an inline-six or a V8. These are examples of this method of description.

Valves

Valves are very similar to doors. They are more precise than the average one will be, however. These parts are very carefully machines to very tight specifications. In the engine, the fuel mixture goes through holes in the cylinder heads known as intake ports. The products of the combustion reaction will then leave the cylinders through exhaust ports. Valves are the parts which govern when these ports are closed and when they are open. Usually there will be one intake valve and one exhaust valve for each cylinder on the vehicle.

A spring holds the valves closed until a part known as the cam makes it open up. Cams are wheels with small bulges on them. A shaft, known as the camshaft, will have a single cam for every valve that needs to be opened. The cam can operate by pushing on the stem itself or through the use of a rocker arm which will do the work for it. Likewise, the rocker arm might utilize a push rod to help it do its work or it might do everything on its own. Somewhere along the length of the area between the cam and the push rod is a component which is called the valve lifter, which is what gets pushed when the cam lobe operates. This is always kept in close proximity to the push rod through the use of hydraulic action.

Lubrication of Engines

Lubrication of the engine is, perhaps, the most essential thing to keeping it functioning properly. Try running an engine with no oil and the engine is very likely going to damage itself beyond repair. Metal is never supposed to directly touch metal in an engine. The moving parts do not truly touch each other. They, instead, sit upon a film of oil which allows them to move freely without causing any damage. The amount of oil is not very thick, and amounts to about the same thickness as paper.

The oil in an engine serves a number of functions. It helps to reduce friction, assists the operation of the engine by acting as a coolant, and it helps with shock absorption. Beyond that, it assists by working as a sealant and a cleaning agent (which is partially why you have to change your oil on a regular basis). Oil accomplishes all of this because of some unique qualities that it has. For one thing, it has a high viscosity, which is the tendency to resist flow (this is why oil is so thick). The viscosity of oil will decrease as the temperature increases

When you buy oil for a vehicle, you may have seen the term SAE. SAE stands for the Society of Automotive Engineers. The number after that indicates how viscous the oil is, with higher numbers being more viscous than lower numbers. The W that can be seen on some containers of oil indicates how viscous the oil is when cold and the latter number indicates how viscous it is when it is hot.

Below is an image showing how oil might move through an engine:

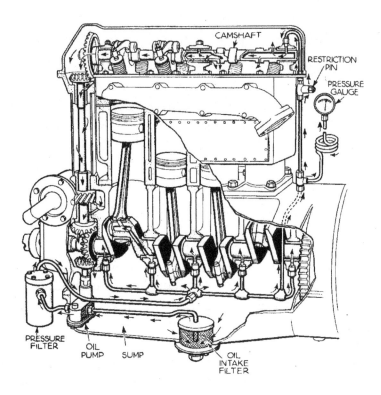

Figure 10.2. Oil movement through an engine

For an engine to function properly, oil pressure must be maintained and the pressure must force the oil to move through the engine and lubricate all of the parts which are going to be moving. The crankshaft helps with this by driving an oil pump which retrieves oil from the oil pan. The oil then moves through a filter and then into a line which will send the oil everywhere that it needs to go.

Oil will flow through any passages available to it in order to get to the moving parts located inside the cylinders and the engine block itself. More than just the cylinders and need to be lubricated by the oil. The valve train requires lubrication as well. Different engines allow oil to lubricate these areas in different ways.

Four Stroke Engines

Four stroke engines work based on a cycle determined by the positions of the piston. The way that the valves operate (closing and opening) have to match the way that the camshaft is moving. The piston moves from either TDC to BDC or vice versa. That movement is known as a stroke. Almost all engines work on a four stroke cycle. The cycle is complete once four strokes have occurred and then the cycle will start over.

- **INTAKE STROKE**: Piston goes down from the TDC position, increasing the volume available in the cylinder and allowing in the fuel and air mixture.
- **COMPRESSION STROKE**: Exhaust and intake valves are going to be closed. The piston will go back to the top of the cylinder and compress the fuel mixture in the cylinder head.
- **POWER STROKE**: Piston begins close to top dead center and then the fuel mixture is ignited and combusts, pushing the piston down to bottom dead center.
- **EXHAUST STROKE**: Piston begins returning to top dead center, forcing the products of the combustion out of the cylinder.

Engine Accessories

The Cooling System

Given the fact that engines have thousands of explosions occurring within them for hours at a time, it should come as no surprise that a high quality and efficient cooling system should be in place. Every combustion reaction that takes place in the engine is capable of producing a temperature which could reach 6,000°F. In other words: Engines get hot. Really hot. A lot of the heat of the combustion reaction will exit the vehicle as exhaust but the heat that remains is going to be absorbed by the various parts of the engine. The oil used to lubricate the engine will cease to function properly once it reaches 450°F, give or take around 50°F. The cylinders, thus, must stay below that temperature.

The engines that are on smaller pieces of equipment have fewer cylinders, usually one or two, and they can be cooled fairly easily simply by the movement of air around them. Engines which are on vehicles, however, must have a dedicated cooling system to accomplish this task. Coolant is a mixture of water and chemicals which is used to help keep engines cool. Usually, the coolant will double as antifreeze, preventing the water from freezing in the engine during times when the temperature drops below freezing. This is moved from the engine inside of tubes called water jackets which are built into the cylinder heads and the engine block. They will then take the water from the engine block into the radiator where it can be cooled. This process also puts the hot water into the heater of the vehicle.

By providing a very large surface area, the radiator lets air move over large portions of piping which cools the coolant that is flowing through it. Using pure water in a radiator, while it does work during emergency situations, is not a good solution to the problem of cooling the engine. It can freeze if the temperature drops low enough and that freezing will damage the cast parts of the engine. Water will also boil at a temperature that is too low to be of much use inside the engine.

Gauges and Lights

The lights and gauges inside of the car are primarily meant to assist you in determining what is going on inside of the engine. The instrument panel has information about the current RPM value of the engine, the oil pressure, engine temperature, and fuel levels. Even beyond that, some modern cars are able to use onboard computers to tell you much more than that (including tire pressure, oxygen levels, and other things).

When you look right past the steering wheel on the interior of most cars you will see the instrument panel. It is here that the gauges are displayed and that warning lights will come on if there is a problem. Typically, if you spot a problem with the engine on your instrument

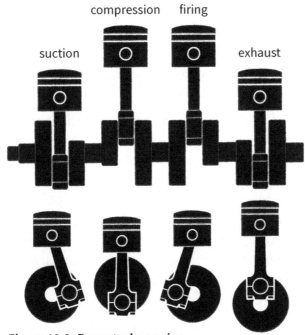

Figure 10.3. Four-stroke engine

panel, you will want to turn the engine off or you run the risk of damaging it. This is also the reason why your engine is at the highest risk of damage when you first turn it on. Since the systems are not primed yet, oil has not been delivered to everything yet. Thus, you should allow the engine to idle for a few seconds before you begin driving (even longer if it is cold outside).

Each cylinder is going to be going through this cycle repeatedly as long as the engine is running.

Fuel and Ignition Systems

The fuel system operates through the use of a fuel and air mixture. When gasoline is in liquid form, it will not burn. What actually burns is the vaporized form of the gasoline that has been mixed with air. This is also why gasoline is so combustible. The process of atomization is when a liquid, in this case the gasoline, is broken down into very small drops which can then be vaporized.

There are two parts in engines which do the vaporization work on the gasoline: fuel injection systems or carburetors. Carburetors have fallen out of use on modern vehicles, but they are still extremely common on many older models. Carburetors, carbs, are mechanical devices which utilize the flow of air to make a vacuum that will pull gasoline from a nozzle in a very fine spray. This does not change the way the fuel flows very quickly if the throttle is being opened or closed rapidly. Due to the position of the carb, it is also not able to provide the maximum efficiency in providing fuel to the cylinders.

Ignition is when the spark plug creates a small spark which ultimately causes the combustion of the fuel mixture inside the cylinder. Spark plugs require around 10,000 volts. That voltage is enough for the current being sent through it to jump from the gap between the two electrodes. The volts are partially boosted by a coil which is located inside of the spark plug. This helps to increase the voltage and decrease the amperage of the electricity flowing through it.

For spark plugs to operate in the way they are designed, they must go off at exactly the right time. Otherwise, their power would go to waste. Gasoline engines usually control the timing of when the spark plugs will go off by using a component known as a distributor. The camshaft of the engine drives the shaft of the distributor. Terminals which are located on the cap of the distributor are connected to the spark plugs by wires and a central terminal is connected to the circuit of the coil. One of the rotor ends is connected to the central terminal and the other ends will line up with side terminals in series. When the rotor spins the shaft voltage surges are sent from the circuit to the spark plug.

It is important that the spark plugs fire at precisely the right time. If a spark plug happens to go off before it is supposed to, then the detonation of the fuel and air mixture can happen. This is called a "knock" in an engine and it is one of the leading causes of engine and piston damage.

Electrical Systems

For a vehicle to function and have power in all of its systems, it has to have ways of both temporarily storing power and routing that power to where it needs to go. The ignition system is not the only part of a vehicle that needs power. Lights, motors, radios, speakers, controllers, and other parts also require the use of power. Most modern vehicles utilize a 12-volt system in order to get power to these systems.

Power can be made in two ways in vehicles. ELECTROMAGNETISM is when a magnet is moved is can create electrical fields around it. When done either around or within a wire, the wire will begin to have a current. How much electricity is produced in this situation depends on how strong the magnet being used is and how quickly it is moving. CHEMICAL REACTIONS also create power. The bonds which keep chemicals together contain energy. When those bonds are broken and when chemicals react with each other, it results in electron flow. This is the primary means through which a battery works.

Batteries are one of the most important parts of modern vehicles. It is from batteries that the electrical current comes from, stemming from a reaction occurring between lead and sulfuric acid inside of the battery. A series of plates of lead alternate with plates of lead oxide in a solution comprised of water and sulfuric acid. These plates are linked to each other through the use of a conductor and separators keep the grids from touching each other. Cells is the term which is typically used to describe the grids inside of battery casings. Each cell makes two volts. To make the twelve volts most cars need, six cells are linked together. More plates equal more electrical power in this case.

Batteries are rated in two ways. AMP HOURS tell how much electricity you can get out of the battery before the chemicals inside are used completely. 50 amp hours would produce 2 amps for 25 hours, 60 would make 3 amps for 20 hours. So on and so forth. COLD CRANKING AMPS tells how much current can come from a battery in 30 seconds at 0°F without dropping below 7.2 volts. This is primarily a measure of how well a battery can hold up in the cold.

The second rating is just as important as the first. Chemical reactions run better when they are hot. In colder temperatures, reactions are not as capable of running efficiently. Because of that, batteries are not able to provide as much amperage as they would otherwise.

A starter is an electrical component which starts the car. The starter serves the function of causing the flywheel on the crankshaft to turn when the ignition switch is turned to the starting position. The flywheel then turns the crankshaft so that it can draw the fuel into the cylinders.

The motor on the starter is capable of rotating the crankshaft at 200 rom and the starter is also the single component that uses the majority of the electricity in the vehicle, requiring 200 amps to function in the right way.

Given the large electrical draw the starter requires, once it has been used you need to have some way to cut it back off. After all, it is only required for the initial start of the engine and it ceases to have a function afterward. Beyond that, the high speeds that the engine will begin rotating can run the risk of damaging or completely destroying the starter. There is a solenoid which is present in the starter which pulls the gear into contact with the flywheel when the starter is being used. When the ignition key for the vehicle is not set to the start position, the solenoid is disengaged and, thus, the current to the starter is removed. A spring then pushes the starter gear back out of contact with the starter.

You may be wondering, at this point, how the battery is able to keep working. Maybe even how it keeps working during while the car is running. The electricity in the battery is being created through the running of chemical reactions inside the battery. If the current is sent to the battery, however, the chemicals inside of the battery will then return to the state they were in to begin with. A device known as an alternator is used inside the vehicle to constantly produce electricity which is then shunted to the battery in order to recharge it. This device is, ultimately, an electromagnet. The magnet is held inside of a coil of wire and is quickly spun by the use of a belt connected to the pulley of the crankshaft. The more rpms the engine is producing the faster the higher the current it creates will be.

Should the alternator not be able to produce enough electricity to recharge the battery, the battery will cease to function (this happens when alternators begin to break down or become damaged). Starting the engine uses a lot of electricity which can quickly drain the battery, but once the car started the engine will produce enough electricity to charge the battery again and replace it.

Transmissions

The transmission is, ultimately, the way that the power produced by the engine is sent to the wheels of the vehicle. The drivetrain is a term which is used to talk about the axles, the transmission itself, and all of the parts which are related to them. A manual transmission is one in which the changing of the gears takes place at the discretion of the driver. Automatic transmissions allow the car and the onboard electronics to determine when the shifting needs to take place. Vehicles that are equipped with a manual transmission will include a clutch.

Rear wheel drive, front wheel drive, and all-wheel drive are terms used to describe where the power is going to move the vehicle. The transmission controls this. On vehicles that are rear wheel drive, the drivetrain will have the transmission, the rear drive axle, and the drive shaft. Front wheel drive vehicles, on the other hand, will have a transaxle that combines the properties of an axle and a transmission. All wheel drive vehicles have both. The most important component that is inside of a transmission is the gears. The turning force required to move the vehicle is known as torque (review your physics). The internal combustion engine is designed to produce a low amount of torque at low engine speeds and, thus, it needs to have the force multiplied to function in the right way.

The principle that applies to this force multiplication is known as leverage. The ratio of the length of the longer end of a level and the pivot point and short end and the pivot point will govern how well the strength of the force can be multiplied. Gears can be considered to be a series of levels on a long shaft. Smaller gears can be used to turn larger gears which can produce more force. A lever is only able to move a certain distance, but a gear can constantly turn around the shaft that it is attached to. Gears can be turned by larger gears. Gears which are doing the turning are the driving gears and the gears which is being turned by the driving gears are known as driven gears.

Figure 10.4. Gears

As you can see in the above image, the "teeth" of one gear fit into another and as the gear turns, the teeth of one push the teeth of another causing the gear to rotate. Gears that are the same size will cause them to turn at the same speed which will have no effect on the torque being produced. Alternatively, a gear that is larger or smaller will change the torque. A driven gear that is twice as large as the driving gear will turn at half the speed. So a driving gear will have to turn twice in order to turn the driven gear (that is twice as large) once. Additionally, the work being done will be affected. This principle is known as gear reduction, because the shaft being driven is turning slower than the one that is doing the driving. Gear reduction is what is used to multiply the torque that can be used by the vehicle. This relationship is described as a gear ratio. In the example used above, there is a 2:1 gear ration. Commonly, a 3:1 gear ratio is used in vehicles.

When the vehicle begins to move at faster speeds, it does not need as much torque to continue that motion (remember the laws of motion). In order to most efficiently utilize the power output of the vehicle's engine, the transmission will use different combinations of gears (usually lower gear ratios) in a way that is proportional to the speed at which the vehicle is moving. When the vehicle "changes gears", you are actually moving the gears physically inside their location to change the gear ratios and modify the torque that the car is getting.

As you can see, when you get right down to it the transmission is primarily just a series of gears that you can control. Different combinations result in different torques and different power outputs. This is usually described by how many gears the transmission has. Four gears would be a four-speed transmission, and so on.

Figure 10.5. Inside of a transmission

Automatic transmissions have an additional gear that manual transmissions do not have, known as the overdrive gear. The ratio of this gear is usually less than one. So it would be 0.8 or 0.6, etc. This helps with lowering the rpm of the engine when cars are traveling at high speeds and, thus, also helps to increase the efficiency of the fuel being used and prevent unnecessary engine wear.

The clutch, if you remember, if the part of the transmission which is located in vehicles utilizing a manual transmission. A clutch consists primarily of a pressure plate, a friction disk, and a flywheel. The flywheel is attached to the pressure plate and the cover of the clutch and the friction disk is located right between them and is attached to the transmission shaft. When these parts are not in direct contact, the clutch will be disengaged and no power is transferred to the drive wheels themselves. When the parts are in contact, the power is delivered to the drive shaft so that the power can be sent to the drive wheels themselves. The clutch is an integral part of manual transmissions.

Automatic transmissions, on the other hand, do not have a clutch. They have a torque converter instead, which is what serves as a "clutch" for that type of transmission. One side has a turbine (the driven member) and is attached to the transmission shaft. The other side is attached to the crankshaft and has a pump (the driving member). The sides do not have direct contact with each other. Transmission fluid is used to fill the assembly and keep it lubricated properly.

Drivetrain

The drivetrain describes how the vehicle powers itself to move. There are three primary methods of drive that these vehicles have: rear-wheel drive, all-wheel drive, and front-wheel drive.

Vehicles typically have the engine and the transmission itself bolted to the front of the vehicle while using their power to drive the wheels located on the rear of the vehicle. When the car is driven from the **REAR WHEELS**, a drive shaft needs to be present in order to transfer power from the transmission to the rear wheels of the car. That drive shaft will move vertically at the end of the axle to allow for a little bit of free movement.

ALL WHEEL DRIVE vehicles are sometimes called 4WD, 4 by 4, or four wheel drive vehicles. The powertrain on these vehicles is able to give power to all of the wheels on the vehicle at the same time. This allows for a powerful driving force able to move these vehicles over rough terrain that would be too much for other vehicles to handle.

Modern vehicles usually operate on a **FRONT WHEEL DRIVE** basis. These have an engine which is mounted sideways. This eliminates the need for the drive shaft and combines the transmission and the drive axle together in one transaxle. The transaxle can be automatic or manual and, when having a manual transmission a clutch will be present.

One interesting thing to note is that the power that the wheels get needs to be modified when going around a corner. This is because when turning corners, the wheels have different distances that they need to travel. The outside wheels have a farther distance to travel than the inside wheels and, if they were going at the same speed, the inside wheel would end up skidding and leading to a lot more wear than was necessary. Thus, the gears inside the rear axle are designed to permit the wheels to spin at different speeds when they are turning or going around corners. Axles have a differential that lets the inner wheels slow down a bit while the outer wheels speed up.

It should be noted that, at the rear axle, another gear reduction is created. The reduction is usually 3:1 (again), give or take 1 reduction. The rear axle ratio, when multiplied by the transmission ratio, is called the final drive ratio. It is important that you choose the right one when you are choosing your combination of gears. Lower ratios will give you good fuel efficiency but will lower the torque and vice versa.

Each type of drive has its own unique advantages when compared to the others. There are disadvantages as well, which bear mentioning in some cases but not in others.

Rear wheel drive sees advantages in the power and design of vehicles of a large size or that will be used for towing. This is the type of drive which is generally preferred when high performing engines are necessary. The downside is that for some applications the steering can be more difficult with this type of drive than most others.

All wheel drive vehicles are good because of increased power, traction, and mobility.

Front wheel drive has an engine located transverse to the transmission. These are extremely common. Without a drive shaft or a rear axle, the weight of these vehicles is much lower than it would otherwise be. The compartment that houses the engine does not have to be as long since the engine itself is turned sideways. Traction can also be improved because of the pulling action of the front wheels of the vehicle rather than the pushing action of the rear wheels.

Chassis and Suspension

The chassis and the suspension are extremely important elements of modern vehicles. These are, ultimately, the parts that control the movement of the body of a vehicle. They assist in the dissipation of the force the vehicle undergoes through movement.

The chassis includes the brakes, the steering column, and the suspension. The brakes are used to slow down the vehicle through the application of pressure to the wheels. Steering columns are utilized to turn the front wheels, with the goal being to modify the direction that the vehicle goes in. The suspension helps to alleviate stress on the vehicle and to help make the ride smoother.

The suspension itself is comprised of a few parts, including multiple types of springs. **SHOCK ABSORBERS** help with handling and help to smooth out the bumps and pitches that come along with driving on uneven terrain. A **TORSION BAR** is a coil spring which has been straightened again. The action of the spring on a torsion bar is due to the bar twisting rather than compressing or stretching. These are common in the suspension of all wheel drive vehicles. A **LEAF SPRING** is a series of layers of metal or plastic strips. When the ends of the spring are compressed, the middle will spring up or down. These return to their original positions once whatever force is acting on the spring dissipates. These are used on rear wheel drive vehicles. COIL SPRINGS are large metal bars which have been coiled up. A coil can be compressed and then return to the original shape that it was prior to compression (or stretching). These are very common and nearly everyone has seen this type of spring.

Shock absorbers might be the most important aspect of a suspension system. They are the component which is primarily responsible for the quick and efficient return of the suspension to the position that it started in. These are what help to govern the way the spring compresses and rebounds by providing fluids through some holes which are located inside the pistons of the shock absorbers. The amount of smoothing that occurs depends on how large the holes in the pistons are. The weight of the shock absorbers and the springs will depend on the size of the vehicle.

The suspension is broken up into a few different parts. The front suspension keeps the front wheel assemblies off of the frame and helps to dissipate the shock from the road. They also help to deal with steering, breakings, and to control the front wheels if they are powered by the drivetrain itself. In vehicles that operate using read wheel drive, coil springs are typically going to be used in the front suspension. These will be mounted on the control arms in order to help the suspension by supporting it. MacPherson struts are becoming more and more popular because of their unique design and the small amount of space that they take up. These will replace the upper control arm on the front suspension. Additionally, they are sometimes utilized in the suspension in the rear of vehicles.

The rear suspension comes in more than one type. It does not have to deal with the steering of the vehicle, which is handled by the front suspension. Solid axle, semi-independent, and independent are the three common types of rear suspensions. Solid axle rear suspensions have the axle suspended along with leaf springs. The axle is capable of moving based on what is happening on the road itself. This is a fairly sturdy design and is flexible enough to work on most consumer vehicles. The issue is that if one wheel hits a big bump, the shock will travel to the other wheel because the axle is one solid piece.

Semi-independent rear suspensions have wheels utilizing a cross member that helps to link the trailing arms in order to assist with general stability of the vehicle. When moving sideways, these types of suspensions utilize the assistance of what is known as a track bar.

Independent rear suspensions are useful on vehicles that are not rear wheel drive. The reason is simple: there is a need for the suspension of front wheel drive vehicles to content with the drivetrain in any way whatsoever. So the rear wheels will have no contact with each other. The rear wheels are mounted on an arm and a swing axle which comes down from the body of the vehicle. These utilize coil springs, shock absorbers, and possible MacPherson struts. This is one of the best types of suspensions simply because forces affecting one wheel will have no effect on any of the other wheels.

Steering, as hard as it might be for you to believe, is one of the most simple and noncomplex systems on a vehicle. The steering wheel is attached to a shaft and when it is turned the shaft will rotate and turn a steering gear. That gear then moves the tie rods which are connected to the arms and knuckles of the steering column. Gear reduction in this gear is what allows for easy turning of the wheels of the vehicle. One of the most common types of steering gears is the rack-and-pinion gear, which is not heavy and allows for a good feel of the steering itself. The wheel is attached to that gear which will then interact with the rack, which is attached to the tie rods which play a role in the turning of the front wheels.

Brake Systems

The brake system is what makes the vehicle stop moving. It would, of course, still come to a stop when it was moving without the brake system, but then the vehicle would have to rely on friction, which will not stop it fast enough to function usefully in society. The friction of the brake material slows the rotation of the wheels when pressure is applied to the brake pedals. The linings which are on the inside of the brakes will press on discs or drums and then heat must be dissipated. This is the reason why some types of brakes will be either vented or finned: to increase the surface area available for heated metal to be in contact with the air.

There are two main types of brake systems that are used on modern vehicles. **Disc brakes** have a disc which moves along with the wheel. The brake linings are then put on a caliper assembly that will allow them to come into contact with that disc in order to slow it, and the wheel, down. **Drum brakes** have an iron or aluminum drum which has been attached to the wheel mounting surface. Brake shoes (metal linings) are then coated with head resistance and friction resistant material and put inside of the drum. When the brakes are applied, the linings will touch the drum and slow it down, braking the vehicle.

Given the fact that two materials have to come into contact with each other at high speeds and the vehicle is slowed down primarily by means of the friction between those two materials, it should come as no surprise that brake systems will wear out over time. The pads and shoes of the brake system are what sees the brunt of the wear. Modern brake systems are capable of repositioning automatically in order to adjust to the loss of material over time. When the sound of metal grinding on metal is heard, it is time to replace the brakes. Hydraulic pressure stemming from the master cylinder is the primary means by which brakes work (except for the parking brake, which is entirely mechanical). The hydraulic pressure is created when the brake pedal is depressed, forcing brake fluid into the lines of the braking system

Figure 10.6. Brake systems

Periodically, just like all other systems, the brake system has to be flushed. This process is when the brake fluid is removed from the system and then new brake fluid is put in. About 1 quart of fluid is going to have to be used to accomplish this process.

Brake systems typically utilize a power booster in order to help reduce the effort required to depress the pedal and to increase the efficiency of the braking itself. These will usually be either vacuum or hydraulic in nature. If there is a hydraulic system, then a hydraulic pump is used to help with the braking assistance. This can use either the power of the engine itself or it can be powered by an entirely separate motor. Vacuum from the engine is used in vacuum-assisted systems.

Measurement Tools

First things first. In the United States, most everything is going to be measured using either feet, indicated by ('), or by inches, indicated by ("). The metric system is utilized here as well, but primarily in scientific settings. For most applications, the foot and the inch are going to be used. Measurement devices are tools which are used to assist with this measurement. Most of these will have feet and inches in addition to centimeters and millimeters, just in case they are required. The inches on many measurement tools which are utilized in the field of woodworking will have the inches divided into smaller increments, such as sixteenths, eighths, quarters, and halves. As the saying goes, "measure twice, cut once". Measurement tools are meant to ensure that your final product turns out the way that the design looks.

Fastening Techniques

Most wood is attached to other pieces of wood through the use of fasteners. There are many types of fasteners on the market, but the most common ones are nails and screws.

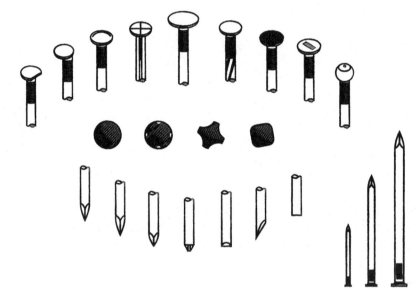

Figure 10.7. Common nails

Fasteners

Most fasteners are going to come in the form of nails. These provide one of the easiest and most effective ways of attaching pieces of wood together and are commonly used in rough projects and those which are inexpensive. This might include things such as crates, boxes, or the framing of a house, among some other things.

Here are descriptions of some of the most common nails you are likely to encounter:

- **COMMON WIRE NAILS**: Large, flat heads. These are different lengths, all of which are heavier than some other types of nails. These are the most common types of nails.
- **BOX NAILS**: Heavy heads with small gauges. These are easy to remove from wood but they also do not hold it as well as some other types of nails can.
- **BRADS**: Brads are typically used for interior trimming and fine projects. These are typically between 3/8 and 3 inches in size,
- **TACKS**: Tacks are often utilized in carpet or upholstery. They have large heads, very sharp points, and they are usually made of iron.

- **BRASS NAILS:** These are made of brass or copper. Usually, they will have tiny heads.
- **SHINGLE, ROOFING, OR PLASTERBOARD NAILS:** These are utilized in special situations in mind and will be primarily used to help prevent any material from shifting.
- **CASING NAILS:** These are used for flooring sometimes and they can be countersunk into the wood.
- **FINISHING NAILS:** Finishing nails are used for moderate to fine work when the nails heads should not be visible. These are typically sized and bought the same way that common nails would be.

FASTENING TECHNIQUES

The utilization of nails is a relatively simple procedure, but here are some tips you can follow in order to use them more effectively and avoid any potential problems:

- Drilling a small hole into hardwoods can be extremely useful prior to driving a nail. Just make sure the hole is smaller than the nail itself is.
- Nails will often go along with the grain in the wood. This can cause the nail to go in the wrong direction. Be mindful of this when applying nails as fasteners.
- Sometimes nails will not go straight into the wood. If that is the case, take it out and drive another nail somewhere else close to that location.
- Sometimes a nail will bend when you are hammering it. If this happens, take it out and use a new nail in the same location.
- Don't drive nails close to each other along the grain of the wood. That will split the wood and cause structural damage.

Driving nails is simple:

1. Put the nail where you need to drive it, holding it with your fingers.
2. Tap the head of the nail to drive it in a bit. Be careful.
3. When it is driven enough to stay where it is, move your fingers and drive the nail more vigorously.
4. Repeat as needed. Nails should end up flush with the surface they are being driven into.

If you need to pull a nail:

1. Slide the hammer's claw under the head of the nail. Make sure that the head is caught in the slot that is on the claw.
2. Pull back on the hammer until it is vertical.

3. Slip a block of wood under the hammer head and apply torque until the nail is completely removed. The wood is primarily there to allow additional leverage.

Screws

Screws are a type of fastener which allow you to hold wood (or other materials) together in a more secure way than nails are capable of doing. Screws can be tightened if needed or removed without causing additional damage to the wood. Screws have a higher cost than nails and they are a bit more difficult to put into use as well. There are many different types of screws and they can be made of a number of different materials (including copper, steel, or brass, among others).

Screws are typically going to be classified based on the type of head that they have. Here is a brief overview of the types of common screws (with Phillips and flathead being the most common):

- PHILLIPS: These screws have cross-shaped slots on the top that look like a plus sign. The driver, thus, is able to center itself more easily and the entire process of screwing becomes a bit easier.
- FLATHEAD: These are typically countersunk because the head is not supposed to show once the screw has been driven.
- OVAL HEAD: Typically used to put together hardware (hinges, particularly) to wood. These are countersunk up to the over part at their top.
- ROUNDHEAD: These are meant for use in situations when the head should be showing.
- DRIVE: These need to be driven into the wood prior to being screwed in. Imagine a cross between a screw and a nail.
- HOOKS AND EYES: These are used for situations when things need to be hung from wood or another surface.
- LAG: These are normally square bolted for the head and are driven with the help of a wrench. Typically, these are going to be utilized when heavy wood is being used.

Screws come in a variety of shapes and sized. They come from one-fourth of an inch all the way up to five 5 inches. Usually, a number between zero and twenty-four will be utilized in order to tell the gauge or the diameter of the screw. The bigger the diameter, the bigger the gauge number that will be associated with it.

Which screw you will use largely depends on the situation that you find yourself in. The gauge and the size and length of the screw will depend on the final use of the screw. Does it need to be hidden? Will you be hanging something from it? What is being held together? Does the screw need to be accessed at a later time for some reason?

To use a screw, you will want to pre-drill a hole into the surface of the item you are going to be applying the screw to. This hole should be smaller (and shorter) than the screw itself. Next, you will insert the screw and apply torque with a drill or screwdriver until it goes inside and can be tightened.

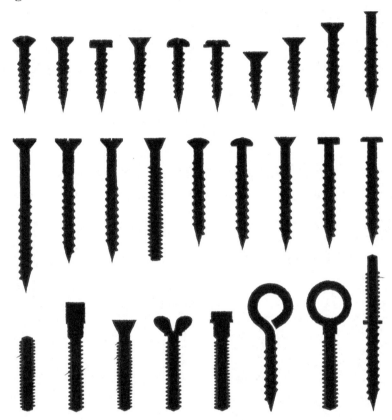

Figure 10.8. Types of screws

Bolts

Bolts are not tapered like screws are. Instead, they have a uniform diameter between both ends. Bolts go through pieces which are meant to be held in contact with each other and then a nut, which is a hexagonal

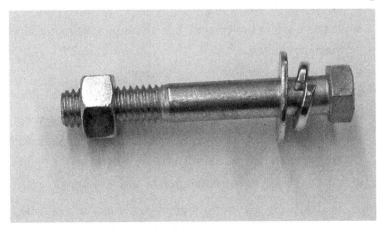

Figure 10.9. Bolt, nut, and washer

or square piece of metal, is attached to the threaded end of the bolt which is projected out. If the bolt is meant to hold something together which can easily be removed (more easily than with a normal nut), a winged nut might be used. Bolts are much stronger than screws or nails and are capable of holding things together with more strength. They are often used in masonry applications or when attaching metals to each other.

Bolts require the presence of a hole that is of uniform diameter with the bolt to be drilled in both of the pieces which will be fastened. A metal washer is then placed between them which helps to spread the pressure of the bolt around on the material, rather than having it in one single location.

Tools

Hand Tools and Other Hardware

Here are some of the most common hand tools that you might encounter:

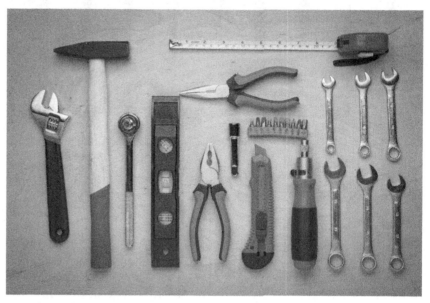

Figure 10.10. Common hand tools

- ◆ **HAMMER:** Used to impact objects and to drive nails.
- ◆ **CHISELS:** Used to remove a small amount of material and to shape things.
- ◆ **SCREWDRIVER:** Used to drive screws
- ◆ **STAPLER:** Used to apply staples as fasteners
- ◆ **WRENCH:** Used to tighten and loosen bolts and screws
- ◆ **SAW:** Used to cut material. Often small amounts, since it is a heavy manual labor burden to use a hand saw for a lot.
- ◆ **PLIERS:** Used to help manipulate and leverage items.

Additionally, there are a number of fasteners you are likely to encounter when doing woodwork. Here are a few of them:

- **HINGES**: Used to connect two objects and allowing a limited amount of rotation of the two objects which are connected.
- **BRACES**: A fastener which is used to clamp things together and provide an amount of support.
- **CORRUGATED FASTENERS**: Thin strips of metals that have alternating grooves which are used to fasten things.

Power Tools

Power tools are used in order to more easily work with materials. Sometimes it would require an extensive amount of manual labor in order to accomplish the same thing that a power tool could accomplish in a fraction of the time. Here are some of the most common power tools and their functions:

A **TABLE OR CIRCULAR SAW** utilizes circular saw blades which are driven by motors. A fence is used to help guide the length that the wood will be cut and the wood is fed into the saw by hand.

A **BAND SAW** is an endless saw blade which runs along a two or three tiered track and is pulled by pulley wheels. It can be used for a variety of materials, but is most often used for wood. Band saws are not as versatile as circular saws are in many cases, but they do have their uses. Band saws should only be operated when the blade is spinning at its highest speed. The stock is fed into the blade by applying light pressure to it. One hand should be used to guide the stock onto the blade. The hands themselves, as with the use of all power tools, should be kept well away from the blade to prevent injury.

The **JIGSAW** is usually used in conjunction with other tools. It is used to help saw outlines and curves. In other words, a jigsaw is used for fine work that cannot be accomplished with the use of a band saw.

A **DRILL PRESS** is a tool which has a vertical column that has been put into a base (or a bench). The top of the column has a motor which causes the drill to work. The motor and the drill are moved up or down in order to drill a hole through the use of a lever which is either hand or foot activated. Presses usually have some sort of mechanism attached to them for the measurement of depth when that is necessary.

A **LATHE** is used to rotate a piece of wood along a horizontal axis so that is can easily be shaped through the use of a hand tool. This usually requires just as much manual and technical skill as it does power. The lathe will see an item placed between the two rotating ends and, as it rotates, a tool such as a chisel will be used to break away parts of the material being manipulated.

ROUTERS are tools which have a motor mounted on a base which is used for cutting curved or straight edges. The router is handheld and then is guided over the stock. The depth can be modified through the use of a depth gauge which is on the unit.

A **JOINTER** is the electric version of a plane. An electric plane is slightly different, but they serve a similar purpose. A jointer is meant to plane the edges of a piece of wood so that those pieces can then be put together. A planar is simply meant to smooth the surface of a piece of wood. The maximum width of the stock that can fit through a joiner is directly proportional to the size of the joiner itself.

SANDERS are used in order to sand down wood and other items to get the desired finish. Typically, two types of sanders are used: disk and belt. A belt sander moves an abrasive belt around and is typically used for the smoothing. A disk sander will have a rotating metal disk which is used to do the smoothing. Usually the disk type of sander can be attached to an electric drill and be used that way as well.

Working with Materials
All About Wood

Wood is used for a huge variety of projects. Different types of wood all have their own unique characteristics which make them good for certain things. The way that the final product will look and feel will also depend on the type of wood that is used. There are two types of wood: hardwoods and softwoods. These two terms actually do not relate to how soft or hard the wood is. Softwoods come from conifers. The most common types that you will encounter here are yellow pine, Douglas fir, hemlock, white pine, western pine, cedar, cypress, redwood, and spruce. Hardwoods come from broadleaf trees (deciduous trees). The most common examples are elm, ash, cherry, walnut, mahogany, chestnut, oak, birch, basswood, maple, and yellow poplar.

Buying lumber is not difficult, but there is some terminology that you will need to understand:

The **GRADE** of the wood is a measure of how many flaws the lumber has. Blemishes are small knots in wood that do not harm the structural integrity. A defect is a damage to the integrity of the wood. Any knot larger than 1.25 inches is going to be a defect.

- A: virtually no defects or blemishes
- B: small defects and blemishes
- C: more defects and blemishes than Grade B
- D: the most defects and blemishes

COMMON LUMBER is not as free of issues as graded lumber usually is, but is less expensive and is good for most general purposes.

- Number 1: small knots but otherwise good wood
- Number 2: large, coarser, defects

- Number 3: a larger number of problems/defects
- Number 4: even more problems/defects
- Number 5: poor lumber; not useful for shop work

Concrete

Concrete is not only one of the most durable and strong materials used for building, but it is also one of the oldest. The Romans were able to create concrete using the materials which were available to them. At its heart, concrete is a mixture of sand, gravel, water, and cement. Mixing them in the right way will create a chemical reaction that results in the compound known as concrete. This is an extremely useful substance for things such as driveways, small steps, and even some buildings. Concrete must be continuously moved or it will set (harden). This is why concrete is often moved in trucks which keep it spinning in the container it is held in.

First, the ingredients are mixed in a trough, wheelbarrow, or bucket. Some concrete comes premixed, which only requires the addition of water. You then mix the ingredients to reach the right consistency. Allowing the mixture to become either too dry or too wet will result in problems with the resulting concrete.

Typically, a frame will be designed using wood into which the concrete can be poured. This form can be in nearly any shape possible. The thickness of the concrete is ultimately going to be dependent on what the intended purpose is. Usually, there will be some tamped down filler which goes underneath the spot where the concrete is going to be laid.

Once the concrete has been poured, you will then want to smooth the top out. Usually, a trowel will be used initially. It will be moved back and forth across it to ensure the concrete is level in a process known as screeding. Using a darby, the last amounts of smoothing will happen to ensure a perfect and level surface along the surfaces of the frame. After smoothing, the concrete is allowed to dry and then a sealer will be applied to prevent anything from getting deep inside the concrete.

When laying bricks or blocks, you will make the same concrete. A mason's trowel is used to put concrete on the bricks and to smooth it out. Once that has been done, the blocks can be put into the proper pattern and then joined using a joiner. It takes concrete about a month to reach its highest levels of strength.

In order to clean concrete that has already been laid, use a brush with stiff bristles and a 5 percent mixture of water and muriatic acid. Always make sure you wear the proper types of protective clothing when you do this, including gloves and eyewear. Afterward, you will use a trisodium phosphate solution in order to flush the wash of the muriatic acid and then rinse it off. A new sealer will need to be applied after this cleaning.

Design

Design is the integral element in every project, whether the worker is aware of that or not. The three most important elements when it comes to design are the usefulness of the intended object, durability, and proportions.

USEFULNESS is another way of describing the purpose for which an object was made. Whatever the reason is that you are making the design is going to determine the usefulness. Additionally, the way it is designed will help determine how well it fulfills that purpose. A ten-feet tall chair is not going to be very useful to anyone and will not accomplish the goals of its creator (insofar as having a chair to use is concerned). DURABILITY is how well an object can hold up through use. This is determined by the materials which are used to make it and how well it is put together. PROPORTIONS is a way of describing the relationship between the parts of an object and the entire object itself. A good relationship in terms of proportion would be 2:3. This relationship (2:3) is not required for good design, but it is a good guideline to move forward from.

The first step in design is a rough sketch, which can then be refined into the document that you use to finally build whatever you are designing. Multiple sketches are usually going to be required to get the design right. The dimensions and other technical information can be added once the idea has been outlined initially.

Objects which are being designed on paper are usually shown from multiple angles. Since paper is only able to present a two-dimensional view of what will eventually be in three dimensions, it is important to make sure that all aspects of it are covered. Usually, a front view will be present on the bottom of the paper, a side view will be beside it. Above that will be a top view, followed by a three-dimensional mockup located above that.

Planning is one of the most important steps in construction. The time you will spend doing this will more than makeup for the time that you save fixing any potential mistakes which could arise. Once the planning is done, you will move forward to the layout. Layout follows a number of steps, but can be outlined by the following: Check for flaws in the lumber, make sure the lumber is the right size and thickness, lay out the general pattern of the lumber, mark lines using a square, measure (and double check) the lumber, saw.

Tips

Try and visualize how multiple components work together. For the shop tools, think about how they would work in the real world. For the automotive section, visualize cars and their various parts working in unison.

Try to come up with some acronyms or rhymes which you can use to help you easily remember the information this section contains.

Don't be tempted to spend all of your time on one question. 99 percent of the time you either know it or you don't. If you don't, then you want to mitigate the risk of running out of time answering the questions that you do have the answers for.

Don't second guess yourself. Again, you either know it or you don't. If you spot and answer and your gut tells you it is right, then pick it and move on. Don't let yourself get tricked.

In all, there are not many tips which will help you with this section. It is all a matter of memorization.

Practice Questions

1. In a vehicle with a manual transmission, what is the component which connects and disconnects the crankshaft from the transmission itself?

    A. assembly pieces

    B. clutch assembly

    C. clutch plate

    D. transmission connecting rod

2. When sharpening some tools, you will need to manually keep their temperature stable. What is this process known as?

    A. sharpening

    B. dipping

    C. tempering

    D. colding

3. What is the purpose of the oil inside of a combustion engine?

    A. act as fuel

    B. make the pistons move

    C. keep the engine hot

    D. lubrication

4. If you need to replace the head on an ax with a broken handle, how would you make the new head fit snugly?

    A. wedge

    B. glue

    C. screws

    D. melted steel

5. How many cylinders would you find in a common combustion engine?

    A. 4

    B. 6

    C. 8

    D. all of the above

6. Which of the following types of rear suspension will provide the smoothest ride that is available?

    A. independent rear suspension

    B. solid axle rear suspension

    C. no rear suspension

    D. semi-independent rear suspension

7. One of the reasons that the MacPherson strut has become popular in modern vehicles is the fact that it:

    A. is cheap

    B. won't break

    C. doesn't take up much space

    D. is extremely heavy, steadying the vehicle

8. Why are some brake systems finned or vented? What is the benefit of drums or discs having these features?

    A. help with cooling

    B. speed up the braking process

    C. help with heating up the brakes

    D. provide lubrication

9. Bolts are usually used to hold some heavy materials together. Usually metal to metal or metal to wood (and sometimes in masonry). What is the purpose of a washer when you are using bolts as fasteners?

    A. to distribute the force

    B. to better hold the materials together

    C. to make the bolt work

    D. to lubricate the nut

10. Braking systems are meant to slow down vehicles. Which are the two common types of braking systems?

    A. drum and bass

    B. disc and drum

    C. bass and kick

    D. friction and metal on metal

11. What are the two most common types of screws you will encounter?

    A. eye and socket

    B. square screw hooks and phillips

    C. Davidson and drive

    D. flat head and phillips head

12. What are the two types of ends on a tack hammer?

    A. power and light work

    B. wood and metal

    C. magnetic and nonmagnetic

    D. metal and graphite

13. What are chisels used for?

    A. remove small bits of material

    B. break things open

    C. hammer in chisel bits

    D. remove sockets

14. What are files used for?

    A. cut wood

    B. cut metal

    C. shaping and smoothing materials

    D. store information

15. What is a common phrase you might hear in the shop?

    A. Measure twice, cut once.

    B. Only metric, no imperial.

    C. Cut twice, measure once.

    D. Measure only in metric.

Automotive and Shop Information Answer Key

1.	B.	9.	A.
2.	C.	10.	B.
3.	D.	11.	D.
4.	A.	12.	C.
5.	D.	13.	A.
6.	A.	14.	C.
7.	C.	15.	A.
8.	A.		

Review

The automotive and shop section of the ASVAB could easily be two separate subtests, but they are rolled into a single one. There is a vast amount of information here which has to be covered. The quickest way to learn how the parts of an engine are arranged is to buy the shop manual for your car and simply go outside and play around with it. Look at how the parts connect with each other and where they go. Don't pull it apart or anything, but try and locate the pieces.

For the shop section, you will just need to memorize it all if you have never worked with hand or power tools before. That is a very niche topic and it is very unlikely you will have enough time to get hands on experience with all of it prior to taking your test.

The **MEASUREMENT** section went over some of the basic terminology involved in the automotive industry. The information contained here is very useful moving forward through the rest of the guide and is right next to essential for doing well on the test itself.

ENGINE BASICS covers the very basics about engines. What the primary components are, how they produce power, and how the engines create the combustion necessary for the car to function properly.

VALVES are used in engines to control intake and exhaust, as well as a number of other applications.

Engines require a large amount of **LUBRICATION** in order to function without breaking. Oil is typically used for this. The lubrication is meant to do a couple of things: keep the heat from breaking the engine and prevent metal parts from destroying each other due to friction.

GAUGES and lights are used for warnings and indications about how the engine is functioning.

Internal combustion engines generate a massive amount of heat. To protect the systems of the vehicle, **COOLING SYSTEMS** must be in place to help dissipate and redistribute that heat.

FOUR STROKE, OR FOUR CYCLE, ENGINES operate on the basis of four individual processes: intake, compression, power, and exhaust.

FUEL AND IGNITION covers the fuel systems that vehicles use to generate force and power and the ignition systems which are used to ignite that fuel for combustion.

The ELECTRICAL SYSTEMS are what transfers and distributes power to a number of small systems in the vehicle including interior lights, radios, etc.

The TRANSMISSION is a series of gears which helps the vehicle to properly distribute power amongst the wheels.

The DRIVETRAIN is the system of components which help to send power to the wheels that will be used for driving.

The CHASSIS AND SUSPENSION are the core parts of the vehicle and the part that absorb much of the shock and impact of moving on rough terrain.

The BRAKES is what is used to slow vehicles down on the road.

MEASUREMENT TOOLS are meant to be used in order to ensure projects are up to specification and are made correctly.

HAND TOOLS are tools requiring the use of manual labor to make them operate as intended.

POWER TOOLS are the tools used when manual labor is just too difficult or time-consuming. They multiply what you are capable of doing by a substantial amount.

WOOD and woodworking are ubiquitous in the field of shop. Many structures are made of wood (not to mention furniture).

CONCRETE is a building material with a wide range of industrial applications.

No project is going to get very far without DESIGN. This covers design basics.

Takeaways

If you can't remember everything, don't start stressing out about it. The important thing is to do as well as you can, not to get a perfect score. If you have never worked on a car before or done anything involving the use of hand tools, it will probably come as no surprise to you that you have trouble with this section. That is not the end of the world. This section is only necessary for those of you wishing to get a job in the military which requires it.

Relax, take your time, and go through this section of the guide one thing at a time. It won't come all at once, so give yourself a break from it and take a step back if you feel yourself getting overwhelmed.

eleven

MECHANICAL COMPREHENSION

Introduction

The mechanical comprehension subtest is meant to help determine how well you understand the basics about physics and forces. This section includes things such as work, pressure, hydraulics, and mechanical advantage. Sometimes, the questions you are asked will include an image that you will need to use to answer.

Study Information

The mechanical comprehension subtest on the ASVAB is an overview of basic physics, how machines work, and how you can use mechanics to your own advantage. This section is not too complicated and does not go too in-depth into the physics, but it is a section which is important to understand for a number of subspecialties. It is not one of the core tests which goes into calculating your AFQT score.

This section of the ASVAB consists of twenty-five questions that need to be answered within nineteen minutes.

Materials

Materials are all of the things which are used for building and constructing thing. Some of these are better for specific uses than others. Wood is good for building small to medium sized structures and for quick projects. Metal is good for precision projects, things that need to have rigorous standards, and for larger structures. Small containers which are meant to hold things might be made of cardboard, paper, or something similar.

Here are some examples of common materials:

- ◆ steel
- ◆ iron
- ◆ cardboard
- ◆ wool

- ◆ wood
- ◆ glass
- ◆ paper
- ◆ cotton

Properties

Different materials have different properties and characteristics. These are important to understand if you want to select the right materials for the right job.

Here are some of the most important material properties:

Table 11.1. Properties of materials

PROPERTY	DEFINITION
Weight	the force that is on an object due to gravity; the amount of force that is required to move the object
Density	a measure of mass per unit of volume; basically how much is inside something
Strength	how well the object can maintain the shape that it has when it is being subjected to pressures and forces from the outside
Contraction	how much the object will shrink when subjected to certain temperatures
Expansion	how much the object will enlarge when subjected to certain temperatures
Absorption	a measure of how well a material can pick up and hold liquid that contacts it
Center of gravity	the point of the object where it can be balanced (equal force on all sides)

Structural Support

Structural support is a concept which combines many of the material properties that were discussed in the previous section. It is, generally, a way to take a specific amount of materials and use it in such a way that it is able to support a given amount of weight. Consider buildings, tables, and other common creations. All of these have to hold up a specific amount of weight, but the amount that they can hold up depends on their specific properties.

Fluid Dynamics

Solids and liquids behave in very different ways. This section will explain what differences exist. Viscosity is a term which is utilized to describe how easily a fluid is able to flow. This is a term which is particularly important when it comes to engines. Another term which is important to understand is compressibility. This is one of the most important concepts in fluid dynamics. Solids can be compressed with relative ease. Liquids, however, cannot be. The pressure of a liquid will spread out amongst the entirety of the liquid and become equal at all points.

The summary of the main points in the study of hydraulics, thus, would be that liquids are very difficult to compress and that pressure is equal throughout a liquid. Pressure, of course, is defined as the force per unit area. Thus, since the pressure is equal everywhere, if you have a small opening at one end of a container, the pressure will be equal to a large opening at the other. So you can use a small force to push on the smaller end and have a large effect on the larger opening.

Mechanical Motion

Mechanical motion is a fancy term for just motion. This is the study of how objects move. There are a few terms you will have to understand before getting started with this, some of which have already been covered in this guide.

Here is a brief refresher. SPEED is the total distance traveled divided by the total time required to travel. VELOCITY is the total displacement divided by the time in which the displacement has taken place. ACCELERATION is the change in speed divided by the time it takes for the change in speed to occur.

You may be having trouble understanding the difference here, so some clarification may be required. Consider the following scenario:

A car drives around a circular track that has a length of two miles. The car drives around the track three times in three minutes. How fast is the car going?

Well, a total of six miles was traveled. Six miles in three minutes is two miles per minute. 2 × 60 = 120, or 120 miles per hour. But the velocity is zero. Why is the velocity zero? Because the car ended exactly where it began. There was no displacement because the track is a circle. This is the primary difference and is a source of confusion. Displacement would be the distance between the ending point and the point that the vehicle started.

Next, you will come to the concept of friction. Two types of friction exist: static and kinetic. Static friction is the type of friction which keeps things from moving. If you are trying to push a heavy crate across the floor and it will not move because of the weight that it carries, you are being prevented from moving due to static friction. Kinetic friction, on the other hand, is the type of friction that slows objects which are already moving. This is why your car will eventually stop moving if you take your foot off the gas.

Engines have a lot of moving parts which have to interact both with other moving parts and with parts of the engine that do not move. The moving parts, during the course of their motion, product kinetic friction. Since the friction is inside the engine, it is called internal friction, which decreases the speed and efficiency of the engine. Oil is used to lubricate engine parts and to help overcome this type of friction.

Centrifugal Force

Most people are familiar with what a centrifuge is. If you aren't, then know that it is a machine which has a design that allows it to spin quickly to separate liquids and solids from each other that have been in solution. The reason this works is relatively simple: the machine is spinning but the liquids inside are trying to continue going in a straight line and, thus, are able to separate based on their relative weights.

Figure 11.1. Centrifuge

The easiest way to think about this is when you are in a car. If you were to take a fast right turn in a vehicle, what would happen? Your body would feel like it was being pulled to the left. In truth, the car is pulling to the right and your body is just trying to maintain the direction that it was already going (in a straight line).

Machines

Simple Machines

Simple machines are utilized in a variety of ways almost universally every single day. These machines are, in truth, something that most people would not even consider machines. To understand how these work, you need to understand the concept of mechanical advantage. Mechanical advantage is a measure of how much a job is being made easier by the assistance of a simple machine.

Here is the formula for calculating mechanical advantage, when R = the distance from the applied force to the pivot point and X = the distance from the pivot point to the magnified force:

$$\text{mechanical advantage} = \frac{R}{X}$$

The first kind of simple machine that you need to understand is the lever. A lever is also referred to as a lever arm. This is going to be a rigid (non-flexible) object which is pivoting around a point. The idea being that force which is applied on one end of the lever will be magnified at the other end.

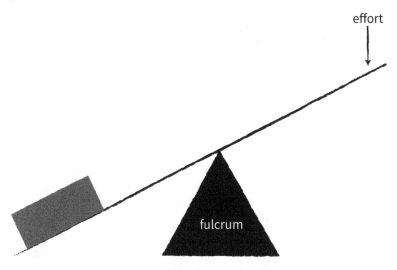

Figure 11.2. Lever

In the image above, when you apply force to the effort arm, the lever will bend about the fulcrum and apply a force to the resistance arm which will be magnified by an amount that can be determined by using the formula for mechanical advantage.

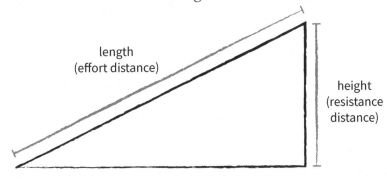

IMA = length ÷ height

Figure 11.3. Inclinced plane

The next type of simple machine that you need to understand is an inclined plane. This is simply a plan which can be used to help you move something from one height to a different height. Usually, these are used to help move heavy boxes or objects from lower points to higher points. Think about this as it is in the real world: a ramp. Pushing a box up a ramp to a height of five feet is a lot easier than picking the box up and lifting it five feet.

The next type of simple machine is a screw. These are typically used to hold things together, often two pieces of wood or metal. The screw has threads which will hold fast to the material and prevent it from being pulled back out easily.

When looking at a screw, think about what is happening when it is going into a material. The threads are removing small bits of material, but only enough to fit itself inside. Thus, it sits snugly inside the material.

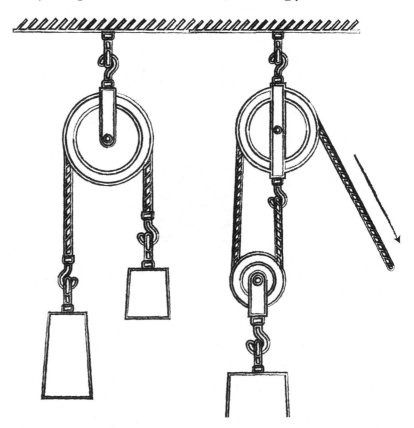

Figure 11.4. Pulley

The next kind of simple machine is a pulley. This is a little bit more complex than the other simple machines that have already been covered. This type of machine has many uses, one of which is helping to lift objects from the ground when used along with a piece of rope.

Pulleys are typically used to pull something from one point to a point at a different height easily. Someone might be using a pulley on the ceiling to move a large box from the floor up to a high shelf, for instance. The applied force can be equal to exerted force in some situations. Depending on the exact way that the pulley is being used, the force that you are applying to the pulley can be greatly increased.

The next type of simple machine is a wedge, which is typically utilized to split objects and separate them from each other. A crowbar might be an example of this (and an example of a lever). Another might be a hatchet or an ax. A nail is an example as well.

Figure 11.5. Wedge

Where the wedge is very thin, it can be inserted and force can be applied to get it to go inside of the object which is to be separated. Once

Figure 11.6. Wheel and axle

that happens, the width of the wedge increases so that the farther it goes into the object, the more it pushes the object apart. In the image above, you can see how this would work.

The next type of simple machine is the wheel and axle. When force is applied to the axle through the turning of the wheel, the axle will transfer the force. The most common example of this is the faucet used to get water outside of your home (or the steering wheel on your car).

When you turn the steering wheel, the axle that is attached to it will turn the gear on the other end, which will, in turn, move the wheels in the directions you want them to go. This is the basis for the steering in vehicles, among many other things.

Compound Machines

Compound machines are machines which have moving parts and more than one component, usually. These are usually utilized for the same general reasons as simple machines would be. They are meant to allow a mechanical advantage for completing a specific task.

A cam is an example of a compound machine. The cam is going be turned by a piston, thus converting the linear motion of the piston into a circular motion. As the rod turns, it turns the ring which is attached to it. This is the method by which many types of devices work, including combustion engines and certain types of pumps.

Another type of compound machine would be a system of gears. One gear will turn and its teeth will then turn a second gear. Small gears will usually have fewer teeth than large gears, and the number of rotations that each gear goes through in the system will change. The mechanical advantage in this situation is defined by how many teeth the gears have. Divide the number of teeth on the large gear by the number of teeth on the small gear and you will have your mechanical advantage. Gear systems are commonly used in vehicles inside of transmissions.

Figure 11.7. Crank

The next type of compound machine is a crank, which is a rod that has a variable radius that has a chain wrapping around the portion of the rod with the larger radius. This is typically used to help lift a weight. The mechanical advantage in this type of machine is the ratio of the large radius of the rod to the small radius of the rod. A colloquial term which is used to describe this type of machine is *winch*.

Another type of compound machine is a linkage, which is a device that is able to convert one type of rotating motion (such as that of a crank) to another type of rotating motion (oscillatory, rotational, or reciprocating). This process, it should be noted, can be reversed. It is not permanent. So you can use a linkage to convert motion in order to turn a crank as well.

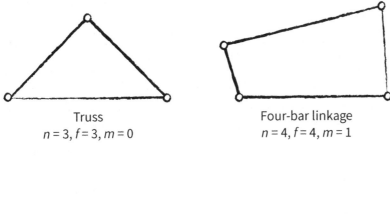

Truss
$n = 3, f = 3, m = 0$

Four-bar linkage
$n = 4, f = 4, m = 1$

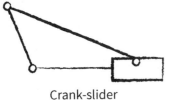

Crank-slider

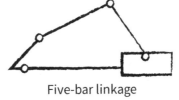

Five-bar linkage

Figure 11.8. Four types of linkages

Tips

Don't spend too much time on any single question. Usually, you will either know how to solve it or you won't. Either way you don't have time to waste, so do everything you can as quickly as you can (while still fully understanding what is being asked of you) and go back to the hard ones later if you have the time.

Narrow down your answers. If you know the answer immediately, mark it down. If you don't, then start ruling out the obviously wrong answers.

Visualize balance when you are being asked questions with regard to structural support. Often, support will be less stringent if the object you are being asked about is unbalanced.

When you are being asked questions about fluids, try to figure out whether or not high or low viscosity would be the better choice for use in the system.

Remember that, when asked about hydraulic systems, the pressure will be equal throughout the system, assuming that the system is of equal height.

Know the difference between speed, acceleration, and velocity. Velocity and speed are terms which are commonly confused, and you can fully expect to be asked a question that will ensure that you know the difference between the two.

One of the fastest ways to begin understanding simple and compound machines is to find examples of them and play around to see how they work. Even if you do not understand all of the mechanics, understanding the basic functionality that is coming into play can be enough to answer many questions on the ASVAB about them.

Go ahead and cross out answers that you find on the test that you already know are incorrect, that way you won't accidentally spend time reconsidering these questions when you already know they are incorrect.

Understand the way that mechanical advantage works for different types of machines. Even though the machines themselves are different, the equation is the same. Even so, it is important that you have a good idea in your head of the different forces that you need to look at.

Understand the different ways that solids and liquids respond to stimuli from the outside. This is the key to understanding materials and their properties. The way they stand up to the environment is the primary way that you determine which one will be used in what situations.

Practice Questions

1. Which of the following provides the best definition for what *strength* would be, in terms of materials?

    A. how much weight it can lift

    B. its ability to maintain shape

    C. how heavy it is

    D. how well it can float

2. Which of the following can cause an object to accelerate?

    A. force

    B. speed

    C. mass

    D. time

3. What is the primary distinguishing factor between speed and velocity?

    A. total movement

    B. space

    C. displacement in time

    D. distance

4. In which of the following materials would you find the highest density?

    A. paper

    B. wood

    C. water

    D. gold

5. What type of energy might you have at the top of a roller coaster?

    A. potential energy

    B. frictional energy

    C. momentum

    D. torque

6. What is the center of gravity of an object?

    A. the heaviest point of the object

    B. the point where the object can be balanced

    C. the most unbalanced point

    D. where gravity affects the object the most

7. What is the definition of viscosity?

    A. how viscous a fluid is

    B. how vicious an object is

    C. the ease with which a fluid can flow

    D. how long before a fluid becomes solid

8. Two footballs have been placed 60 centimeters apart. They are identical in every way. Where is the center of gravity relative to the second football?

    A. 30 centimeters away toward football 1

    B. 25 centimeters away toward football 1

    C. 40 centimeters away toward football 1

    D. 60 centimeters away toward football 1

9. In which situation are you using a mechanical advantage to assist yourself?

    A. pushing your car

    B. picking up an apple

    C. removing a nail using a claw hammer

    D. climbing a ladder

10. When would you not need to use a pump?

    A. move water from the ground to the second floor of a house

    B. move water downhill

    C. move water into a pool slightly uphill

    D. get water out of a large enclosement

11. Which of the following is an example of a simple machine?

    A. a screw

    B. an engine

    C. a radio

    D. a small robot

12. Which of the following is an example of a complex machine?

    A. screw

    B. pulley

    C. incline plane

    D. crank

13. If gear one turns clockwise, which way does gear two turn (assuming they are connected)?

    A. counterclockwise

    B. clockwise

    C. it doesn't turn

    D. not enough information

14. What kind of force is used to slow down a car?

    A. centripetal

    B. centrifugal

    C. friction

    D. gravity

15. An increase in speed is known as:

    A. deceleration

    B. acceleration

    C. force

    D. velocity

GO ON

Mechanical Comprehension
Answer Key

1.	B.	9.	C.
2.	A.	10.	B.
3.	C.	11.	A.
4.	D.	12.	D.
5.	A.	13.	A.
6.	B.	14.	C.
7.	C.	15.	B.
8.	A.		

Review

The mechanical comprehension section of the ASVAB is primarily used to help the military figure out composite scores so they can place you into a specialty. It is utilized to ensure that you are able to understand how simple and slightly complex machines work, how to utilize forces involved in basic physics, and to employ mechanical advantage when it is of benefit to you to do so. They may show an image and then ask you questions about in on the test in order to make sure you adequately understand specific concepts.

The **MATERIALS** section discusses different types of materials, the properties of those materials, when they might be used, etc.

STRUCTURAL SUPPORT is the primary means by which weight can be held up by a given structure.

FLUID DYNAMICS is how fluids work. This also covers information about the basics of hydraulics.

MECHANICAL MOTION is about how movement works. This includes forces, speed, acceleration, velocity, and information about hydraulics.

CENTRIFUGAL MOTION is a type of motion which is used when you need to separate solids from liquids or to separate different substances based on their relative densities.

SIMPLE MACHINES are simple, one part machines. Information about screws, planes, wedges, and pulleys, among others.

COMPOUND MACHINES are machines with moving parts and multiple components.

Takeaways

The entire point of this subtest is to make sure you understand three primary things: how forces interact with objects, how work is done on objects, and how machines work. If you have a broad understanding of these three topics, then you should have no issue with the mechanical comprehension portion of the ASVAB. Though the machines which are covered here are a bit complicated at times, the basic way that mechanical advantage works does not change. It is important that you understand the concepts here, at least on a narrow and shallow level, to do well on this portion of the ASVAB. This is especially true when it comes to both hydraulics and mechanical advantage, which are one of the tests favorite types of questions.

twelve

ASSEMBLING OBJECTS

The Assembling Objects section is a bit different from the rest. You cannot study for it like you do for math, reading, or science. It calls for a set of skills that some people just understand better than ohters. In fact, many recruiters will advise that you not even complete this section; but since it won't hurt to do so, we'll include a brief review.

There is a sixteen-minute time limit to answer sixteen questions of which there are two types. The first will require you to match up points on objects as indicated. The second is similar to a puzzle: connect various shapes together to form one cohesive unit. The easiest way to solve both of these types of questions is through the process of elimination. Look at those answer choices which cannot possibly be correct. Perhaps they do not have the same points as those indicated in the question, or maybe they have different shapes than those listed in the question. Eliminate those, and you'll only have one or two selections that can be correct. From there, you can reason your way to the right answer!

The best way to understand these concepts is through practice - but don't spend too much time on this section. The others are a much higher priority.

Practice Questions

1.

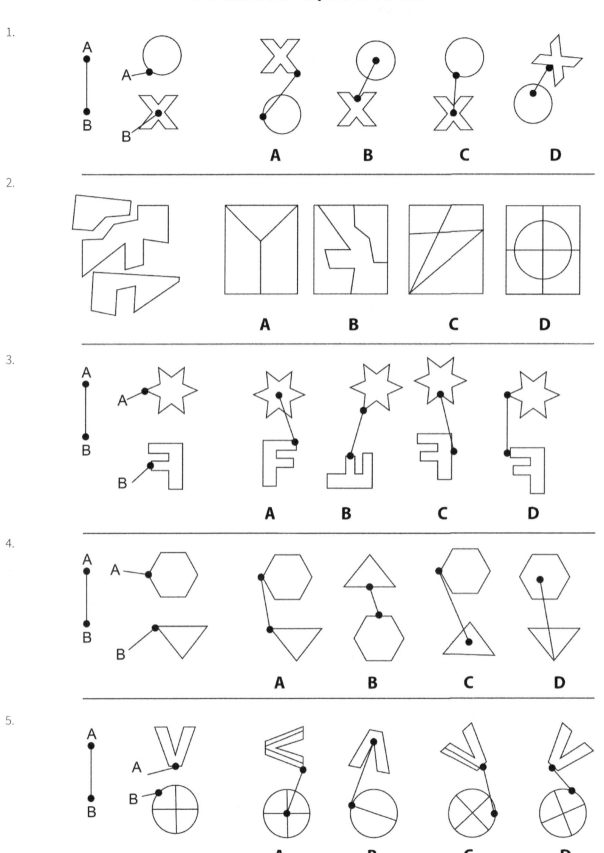

2.

3.

4.

5.

6.

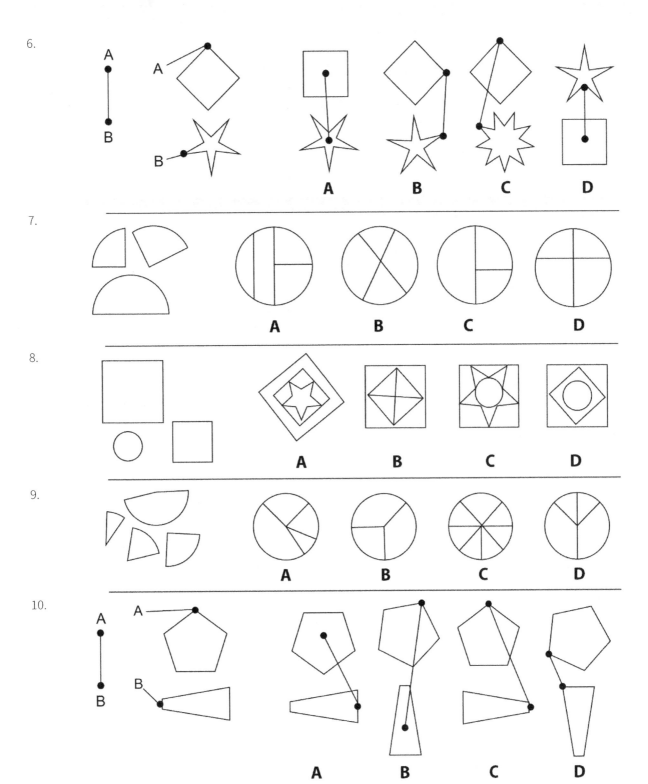

7.

8.

9.

10.

Assembling Objects Answer Key

1.	B.	6.	B.
2.	C.	7.	C.
3.	D.	8.	D.
4.	C.	9.	A.
5.	A.	10.	C.

PART III: TEST YOUR KNOWLEDGE

Introduction

The following practice exam will take you through all of the individual sections of the ASVAB subtest by subtest. To get the most out of this, you should have (ideally) already worked your way through the rest of the guide. Hopefully, you took a practice exam prior to beginning the guide as well (that way you will have a benchmark). Consult the introductory section of this guide and time yourself according to the time limits that have been laid out there.

By completing this practice test, you will have a good idea of where you stand when it comes to the actual ASVAB. If you don't do well, do not beat yourself up. Studying for the ASVAB takes a significant amount of time and effort, and every step that you take toward your goal is another brick on the road to success.

Again, here is a breakdown of the test:

Table 13.1. Paper and pencil ASVAB content

SUBTEST	TIME LIMIT (MINUTES)	NUMBER OF QUESTIONS
General Science (GS)	11	25
Arithmetic Reasoning (AR)	36	30
Word Knowledge (WK)	11	35
Paragraph Comprehension (PC)	13	15
Mathematics Knowledge (MK)	24	25
Electronics Information (EI)	9	20
Auto and Shop Information (AS)	11	25
Mechanical Comprehension (MC)	19	25
Assembling Objects (AO)	15	25
Totals	**149**	**225**

Here are the steps you should take before taking this practice exam:

1. Review the general outline of the test, take note of the time limits here and the number of questions for each subtest.
2. Divide the time limit for each subtest by the number of questions for that same subtest to get your average time per question.
3. Create your answer sheet, divided by subtests and numbered.
4. Get your pencils, scratch paper, timer, and erasers ready.
5. Begin the test.

thirteen

PRACTICE TEST ONE

General Science

1. What is the basic unit of volume in the metric system?

    A. meter

    B. milliliter

    C. liter

    D. gram

2. What is the field of geology primarily concerned with?

    A. the Earth

    B. volcanos

    C. the sky

    D. the moon

3. There are two primarily building blocks which are used by plants during photosynthesis in order to make sugars (primarily glucose). What are they?

    A. water and carbon dioxide

    B. dirt and air

    C. nitrogen and carbon

    D. water and oxygen

4. What is the fundamental biological unit that is used in order to build protein structures?

    A. nucleic acids

    B. amino acids

    C. elements

    D. water

5. What is chemistry?

    A. study of chemical elements

    B. study of the earth

    C. study of the planets

    D. the study of life

6. What is the organ of the human body which helps to pump blood throughout the circulatory system?

    A. lungs

    B. heart

    C. hemoglobin

    D. brain

7. In 1 inch, there are 2.54 centimeters. In 1 foot, there are 12 inches. How many centimeters are there in 1 foot?

    A. 30.48

    B. 10.43

    C. 12

    D. 24.5

8. The outermost part of a cell is known as the plasma membrane. What is the plasma membrane called in plants and bacteria (and some fungi)?

    A. plasma membrane

    B. permeable membrane

    C. cell wall

    D. mosaic membrane

9. What is one method to convert the Celsius scale to the Fahrenheit scale?

    A. $C = F + 32$

    B. $C = 32 + \left(\frac{9}{5}\right)F$

    C. $F = 32 + C\left(\frac{5}{9}\right)$

    D. $C = 95 \times F$

10. Which organ is the location at which circulating blood is oxygenated in order to carry oxygen to the tissues of the body?

    A. brain

    B. heart

    C. lungs

    D. skin

11. Which kind of subatomic particle is found outside of the nucleus (and carries a negative charge)?

    A. electronino

    B. neutron

    C. proton

    D. electron

12. What is absolute zero?

    A. 0°C

    B. 0 Kelvin

    C. −32°C

    D. 0°F

13 The process of cell division which results in two daughter cells with the same number of chromosomes as the parent cell is _____.

    A. mitosis

    B. division

    C. meiosis

    D. sex

14. Which subatomic particle is used to determine the atomic number of an element?

    A. neutron

    B. proton

    C. neutrino

    D. electron

15. What is the term that is primarily used for a gene which will be expressed even when a different allele is present?

    A. pragmatic

    B. recessive

    C. dominant

    D. denim

16. _____ is the process through which cells convert the nutrients that they received into ATP.

    A. cellular respiration

    B. diffusion

    C. permeability

    D. nuclear splitting

17. What is the highest (broadest) level of classification in biology?

 A. phylum

 B. species

 C. class

 D. kingdom

18. What is the first element on the periodic table?

 A. plutonium

 B. iron

 C. hydrogen

 D. helium

19. What color of light is not used by the leaves of most plants?

 A. red

 B. blue

 C. violet

 D. green

20. What is the definition of force?

 A. weight due to gravity

 B. the color of an object

 C. speed across a distance

 D. interaction resulting in a change of motion

21. What type of energy is used by plants during the process of photosynthesis in order to assist in making sugar?

 A. moonlight

 B. sunlight

 C. electricity

 D. friction

22. What is the study of physics concerned with?

 A. matter and its motion

 B. chemical substances

 C. life processes

 D. telekinesis

23. _____ is a term which is used to describe the changes in organisms over a period of time in terms of their traits and genes.

 A. chemistry

 B. biology

 C. evolution

 D. astronomy

24. What is the basic unit of distance in the metric system?

 A. liter

 B. meter

 C. gram

 D. centimeter

25. What is genetic drift?

 A. when a group of individuals leaves a large population and becomes genetically different over time

 B. when genes float through the bloodstream

 C. when seeds float on air currents

 D. when biological "rafts" carry cells to new locations

GO ON

Arithmetic Reasoning

1. If the total sum of the ages of William, Scott, and Jim is 80 years, then what was the sum of their ages five years ago?

   A) 75 years
   B) 65 years
   C) 50 years
   D) 69 years

2. A company is organizing a party for one of their employees who is retiring. One-fifth of the women in the office decided to attend along with one-eighth of the men. What is the fraction of the total number of employees who attended the retirement party?

   A) $\frac{5}{13}$
   B) $\frac{26}{80}$
   C) Not enough information
   D) None of these answers

3. Some friends are going to go to a potluck. Everyone has decided to spend a total sum of $96 on the food they are bringing. Four of the people who were supposed to attend did not show. Because of their absence, everyone had to kick in an extra $4 to make up the difference in food cost. How many people attended the potluck total?

   A) 96
   B) 14
   C) 20
   D) 8

4. Bags of apples cost exactly $7 apiece. Bags of pears cost exactly $5 apiece. Jim goes to the store and spends a total of exactly $38. How many bags of pears did he buy?

   A) 1
   B) 2
   C) 3
   D) 4

5. Select the next 4 numbers in the following series: (2, 4, 6, 8...)

   A) 10, 12, 14, 16
   B) 9, 10, 11, 12
   C) 1, 3, 5, 7
   D) 12, 14, 16, 18

6. John operates a small saw mill. Out of the workers at the mill, one-fourth of them are female, and one-eighth of the females are from out of town. What proportion of the mill workers would you expect to be both female and from out of town?

   A) $\frac{1}{4}$
   B) $\frac{1}{5}$
   C) $\frac{1}{8}$
   D) $\frac{1}{40}$

7. There is a hospital that operates in the town, employing 214 total employees. Out of those 214, 63 of them are male. What percentage of the employees of the hospital are male, rounded to the nearest percent?

   A) 10%
   B) 30%
   C) 29.9%
   D) 28.8%

8. What is the product of 87 and 92, rounded to the nearest ten?

   A) 8,004
   B) 8,000
   C) 8,800
   D) 7,800

9. Patients who are severely burned need to have treatment quickly. Once one hour has passed, the chance that the patient will survive drops at a rate of around 12% for every hour. If the individual does not get treatment for six hours, what is their chance of survival (choose the closest percentage given)?

A) 72%

B) 60%

C) 40%

D) 28%

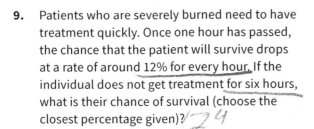

10. John is trying to figure out what his average test grade is for the semester. He knows that the lowest grade he received on a test will be dropped and will not be used to figure out the average. His scores are 64, 21, 76, 80, and 85. What is his average?

A) 76.25%

B) 76%

C) 65%

D) 65.2%

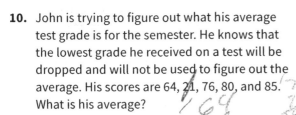

11. Jane earns $15 per hour babysitting. If she starts out with $275 in her bank account, which of the following equations represents how many hours will she have to babysit for her account to reach $400?

A) $-400 = 15h - 275$

B) $400 = \frac{15}{h} + 275$

C) $400 = 15h$

D) $400 = 15h + 275$

12. If a map has a scale of $\frac{1}{8}$ of an inch equaling 5 miles, and the map is 12 inches, how many miles across does the map cover?

A) 400

B) 1200

C) 4800

D) 480

13. In a pack of 100 dogs, 86% of them are female. How many dogs are female?

A) 8

B) 86

C) 860

D) 8.6

14. A rest home in the local area has 56 males and 93 females. Out of the residents, what is the percentage of male residents?

A) 56%

B) 37%

C) 37.5%

D) 149

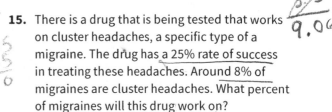

15. There is a drug that is being tested that works on cluster headaches, a specific type of a migraine. The drug has a 25% rate of success in treating these headaches. Around 8% of migraines are cluster headaches. What percent of migraines will this drug work on?

A) 25%

B) 8%

C) 22%

D) 2%

16. At a bake sale, muffins are priced at $1.50 each and cookies are priced at $1 for two. If 11 muffins were sold, and the total money earned was $29.50, how many cookies were sold?

A) 12

B) 13

C) 23

D) 26

17. The post office is having trouble figuring out their staffing. They need to have one person deliver mail for every 183 houses in the town. The town has 2,984 houses. How many mail carriers will have to be hired?

A) 16.3

B) 17

C) 18

D) 20

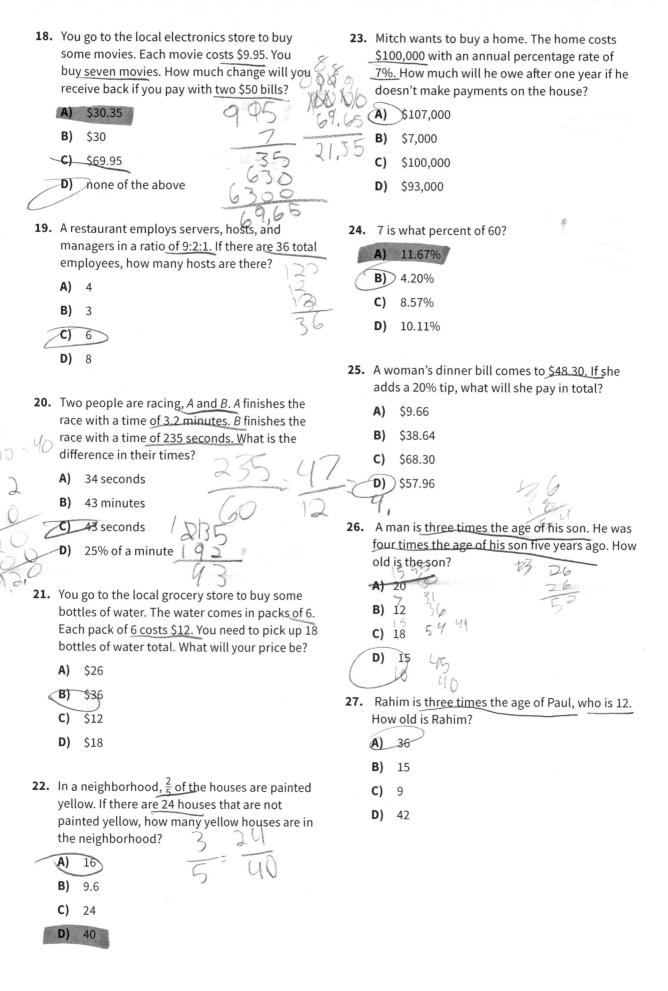

18. You go to the local electronics store to buy some movies. Each movie costs $9.95. You buy seven movies. How much change will you receive back if you pay with two $50 bills?

    A) $30.35

    B) $30

    C) $69.95

    D)   none of the above

19. A restaurant employs servers, hosts, and managers in a ratio of 9:2:1. If there are 36 total employees, how many hosts are there?

    A)   4

    B)   3

    C)   6

    D)   8

20. Two people are racing, *A* and *B*. *A* finishes the race with a time of 3.2 minutes. *B* finishes the race with a time of 235 seconds. What is the difference in their times?

    A)   34 seconds

    B)   43 minutes

    C)   43 seconds

    D)   25% of a minute

21. You go to the local grocery store to buy some bottles of water. The water comes in packs of 6. Each pack of 6 costs $12. You need to pick up 18 bottles of water total. What will your price be?

    A) $26

    B) $36

    C) $12

    D) $18

22. In a neighborhood, $\frac{2}{5}$ of the houses are painted yellow. If there are 24 houses that are not painted yellow, how many yellow houses are in the neighborhood?

    A)   16

    B)   9.6

    C)   24

    D)   40

23. Mitch wants to buy a home. The home costs $100,000 with an annual percentage rate of 7%. How much will he owe after one year if he doesn't make payments on the house?

    A) $107,000

    B) $7,000

    C) $100,000

    D) $93,000

24. 7 is what percent of 60?

    A)   11.67%

    B)   4.20%

    C)   8.57%

    D)   10.11%

25. A woman's dinner bill comes to $48.30. If she adds a 20% tip, what will she pay in total?

    A) $9.66

    B) $38.64

    C) $68.30

    D) $57.96

26. A man is three times the age of his son. He was four times the age of his son five years ago. How old is the son?

    A)   20

    B)   12

    C)   18

    D)   15

27. Rahim is three times the age of Paul, who is 12. How old is Rahim?

    A)   36

    B)   15

    C)   9

    D)   42

28. How many days does it take 4 cats to kill 4 mice? 100 cats are able to kill 100 mice in a period of 100 days.

A) 100 days

B) 4 days

C) 10 days

D) 1 day

29. Josh, Alex, and Drake have ages that total to 75 years. What was that total two years ago?

A) 69 years

B) 81 years

C) 6 years

D) 25 years

30. Jim is working on a project. He is completing 12% of the original amount of work per hour. How many hours will it take him to finish the project to completion (rounded up to the nearest hour)?

A) 8 hours

B) 8.3 hours

C) 9 hours

D) 0.7 hours

GO ON

Word Knowledge

1. _____ is a word that is a synonym of hardship.
    A. ease
    B. corrosive
    C. care
    D. hardship

2. What is the definition of the word grudging?
    A. rocky
    B. smoothing
    C. not generous
    D. close

3. The parents were afraid the children were going to mar the living room furniture.
    A. damage
    B. fix
    C. move
    D. climb on

4. What is a scabbard?
    A. a painting
    B. the hook to hang a clock
    C. an Arabic boat
    D. a sheath used for a sword or dagger

5. "She upbraided her husband when he came in with mud on his boots." Upbraid most closely means:
    A. scold
    B. hug
    C. love
    D. hate

6. Warble means:
    A. ogle
    B. quaver
    C. cross
    D. drown

7. Which of the following is a possible definition of the word minute?
    A. an opinion
    B. a stain
    C. immeasurably small
    D. very large

8. The French signed the peace accord to end the war.
    A. an argument
    B. a treaty
    C. a piece of paper
    D. a scroll

9. Which of the following words is most closely an antonym of evident?
    A. clear
    B. angry
    C. hidden
    D. in the air

10. What does fraught mean?
    A. cute
    B. fast
    C. worried
    D. rare

11. The man intended to go to work, but he ended up at the movies instead.
    A. planned
    B. did not want
    C. loved
    D. hated

12. Conceit means:
    A. a type of jacket
    B. slowly
    C. quickly
    D. arrogant

13. The _____ student was smart enough to learn, but did not want to, and he was put into the corner to think about what he had done.

   A. apt

   B. refractory

   C. victorious

   D. foolish

14. Which of the following words most closely matches the definition of the word equable?

   A. awesome

   B. dead

   C. rotten

   D. temperate

15. If someone is devoted to the care of sheep or some sort of cattle, you might describe their life as:

   A. farm-heavy

   B. boring

   C. pastoral

   D. leaden

16. The crying woman in the field had received the news in a way unintended by the messenger, and she experienced clear _____, bordering on alarm.

   A. disapprobation

   B. anger

   C. happiness

   D. joy

17. If you were to omit or suppress part of a word or a sentence, you have made an:

   A. error

   B. ellipsis

   C. choice

   D. mistake

18. Pusillanimous most closely means:

   A. loving

   B. cowardly

   C. strong

   D. putrid

19. What is a yeoman?

   A. the farmer who works land that he owns

   B. slave

   C. market

   D. yak owner

20. What is a precept?

   A. A teacher

   B. A representative

   C. Someone who came first

   D. A doctrine being taught

21. The current state of surveillance in the United States is something which is being maintained by a _____ of corporate interests and government agencies.

   A. anger

   B. friendship

   C. nexus

   D. trickery

22. A type of legislation which is used to help ingratiate representatives with the constituents that they have under them is called:

   A. bacon grease

   B. pork barrel

   C. animal bucket

   D. filibuster

23. She ensconced herself in the lounge chair.

   A. fix firmly

   B. sleep

   C. fall

   D. feel sick

24. What is a synonym of punctually?

   A. late
   B. duly
   C. too fast
   D. punctured

25. What is a vicissitude?

   A. a type of intestinal disorder
   B. many-legged insect
   C. a change in circumstance
   D. a strong wind

26. Pendulous is a word which means _____.

   A. strong
   B. tough
   C. stalwart
   D. drooping

27. John was a bibliophile, choosing to spend the majority of his time in libraries.

   A. loved books
   B. hated books
   C. loved libraries
   D. loves to read

28. What is a nicety?

   A. mean word
   B. nuance
   C. open statement
   D. proxy

29. If someone is undergoing privation, they are:

   A. celebrities
   B. private people
   C. rich
   D. lacking necessities

30. Superannuated is a term that means:

   A. brand new
   B. old
   C. born last year
   D. tough

31. Any fact which has been well established throughout the course of history cannot be gainsaid easily.

   A. loved
   B. written about
   C. taken exception to
   D. accepted

32. What does it mean to execrate?

   A. to love unconditionally
   B. to run away quickly
   C. to defecate
   D. to curse or despise

33. What is one possible definition of the word conceit?

   A. not-permitted
   B. tight
   C. turn of phrase
   D. grouped closely

34. What is the term for a closed meeting of members of the same political party?

   A. election
   B. caucus
   C. meeting
   D. vote

35. What does I it mean if you yell in a stentorian way?

   A. quietly
   B. with a booming voice
   C. silently
   D. with anger

Paragraph Comprehension

Prompt One

Young Conrad's birthday was fixed for his espousals. The company was assembled in the chapel of the Castle, and everything ready for beginning the divine office, when Conrad himself was missing. Manfred, impatient of the least delay, and who had not observed his son retire, despatched one of his attendants to summon the young Prince. The servant, who had not stayed long enough to have crossed the court to Conrad's apartment, came running back breathless, in a frantic manner, his eyes staring, and foaming at the month. He said nothing, but pointed to the court.

The Castle of Otranto by Horace Walpole

1. What is the general mood of this passage?

    A. happy

    B. depressing

    C. frantic

    D. hopeful

2. On which day was Conrad to be married?

    A. the birthday of his wife

    B. his own birthday

    C. his father's birthday

    D. the day after his birthday

Prompt Two

In the past, many cars were a manual transmission. Today, however, cars have shifted over to automatic transmission (for the most part). Shifting gears in a manual, however, is an important skill to learn if you plan to hit the road. Simply depress the clutch and then shift with the shifting lever to get the right gear. Then release the clutch and apply pressure to the gas at the same time.

3. Why have cars shifted from manual to automatic transmissions?

    A. because manuals no longer work

    B. manuals are too complex

    C. to lower costs

    D. not enough information

5. What kind of transmission do most modern cars have?

    A. automatic

    B. manual

    C. shifting

    D. auto gear

4. What is the second step in shifting gears in a manual transmission?

    A. press the gas

    B. press the clutch

    C. move the shifting lever

    D. press the brake

GO ON

Prompt Three

These visions faded when I perused, for the first time, those poets whose effusions entranced my soul and lifted it to heaven. I also became a poet and for one year lived in a paradise of my own creation; I imagined that I also might obtain a niche in the temple where the names of Homer and Shakespeare are consecrated. You are well acquainted with my failure and how heavily I bore the disappointment. But just at that time I inherited the fortune of my cousin, and my thoughts were turned into the channel of their earlier bent.

Frankenstein by Mary Shelley

6. Why did the narrator stop having visions?

   A. He discovered poetry.

   B. He died.

   C. He went to heaven.

   D. His soul was lost.

7. Where did the narrator live after becoming a poet?

   A. his house

   B. a paradise of his own creation

   C. a temple

   D. none of the above

Prompt Four

Jim was going to the store to buy apples when he was sidetracked. Sally had been following him the entire time and finally decided to call out. Jim has broken up with her for a reason, and it was ridiculous to think she was still trying to get his attention.

8. Why might Jim not be happy to see Sally?

   A. he is too busy to talk to her

   B. she hates apples

   C. they broke up

   D. she hates him

Prompt Five

The House of Representatives shall be composed of Members chosen every second Year by the People of the several States, and the Electors in each State shall have the Qualifications requisite for Electors of the most numerous Branch of the State Legislature.

The United States Constitution

9. How often are the members of the House of Representatives elected?

   A. every 4 years

   B. every 3 years

   C. every year

   D. every 2 years

Prompt Six

When in the Course of human events, it becomes necessary for one people to dissolve the political bands which have connected them with another, and to assume among the powers of the earth, the separate and equal station to which the Laws of Nature and of Nature's God entitle them, a decent respect to the opinions of mankind requires that they should declare the causes which impel them to the separation.

The Declaration of Independence

10. What is this prompt introducing?

A. the reasons for a separation

B. reasons to stay together

C. a revolution

D. human history

11. Which of the following might mean the same as *dissolve political bands*?

A. make a treaty

B. get rid of the government

C. abolish slavery

D. move away

Prompt Seven

Vampires are known to be wary of men who have, on their person, garlic, crosses, holy water, or bibles. They tend to steer clear of these men, as they see them as dangerous to their continued existence.

12. Which of the following do vampires avoid?

A. garlic

B. holy water

C. crosses

D. all of the above

Prompt Eight

No Senator or Representative shall, during the Time for which he was elected, be appointed to any civil Office under the Authority of the United States, which shall have been created, or the Emoluments whereof shall have been encreased during such time; and no Person holding any Office under the United States, shall be a Member of either House during his Continuance in Office.

The United States Constitution

13. What is this meant to state?

A. Representatives cannot create job for themselves and give themselves those jobs

B. Representatives cannot be paid

C. Representatives cannot be civil servants

D. Representatives must quit their jobs

GO ON

Prompt Nine

Today was not a good day. It all started with the rain in the morning. The windows were down on the car, so the seats got all wet. Then the call from Juliet, and the breakup. After that, I lost my job. Today was not a good day at all.

14. What was the last sign that *today was not a good day*?

    A. rain

    B. car seat

    C. call from Juliet

    D. lost job

Prompt 10:

When Dr. Van Helsing and Dr. Seward had come back from seeing poor Renfield, we went gravely into what was to be done. First, Dr. Seward told us that when he and Dr. Van Helsing had gone down to the room below they had found Renfield lying on the floor, all in a heap. His face was all bruised and crushed in, and the bones of the neck were broken.

Dracula by Bram Stoker

15. What does the narrator mean by *went gravely into what was to be done*?

    A. kill each other

    B. go to a grave

    C. dig a grave

    D. make a plan

Mathematics Knowledge

12.3'
3.41

1. Simplify the expression $(5x^2)^{10}$.

$5^{10}x^{12}$

   A. $5^{10}x^2$

   B. $(5x)^{12}$

   C. $5x^{-8}$

   D. $(50x)^2$

6. If $f(x) = |x - 28|$, evaluate $f(-12)$.

   A. -16

   B. 40

   C. 16

   D. -40

$-12-28$

1023
.6820
- 40 34100
41.943

2. Evaluate the expression $\frac{x^2 - 2y}{y}$ when $x = 20$ and $y = 10$.

   A. 0

   B. 38

   C. 36

   D. 19

$20^2 - 2(10)$

$40 - \frac{10}{20} = 40-2$

$\frac{10}{}$

7. $\frac{10^8}{10^3} =$

17.83
3.41

   A. 10^5

   B. 10^6

   C. 10^{11}

   D. 10^{10}

2728
23870
34100
60698

4.3.196

3. Adam is painting the outside of a 4-walled shed. The shed is 5 feet wide, 4 feet deep, and 7 feet high. How much paint will Adam need, if he includes the top of the shed?

   A. 126 ft^2

   B. 140 ft^3

   C. 63 ft^2

   D. 46 feet

235
35
20
28
126

8. A circular swimming pool has a circumference of 49 feet. What is the diameter of the pool?

   A. 15.6 feet

   B. 12.3 feet

   C. 7.8 feet

   D. 17.8 feet

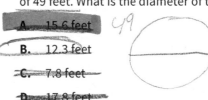

49
2πr
πd = 49
d = 3.41
15.6²
3.41

4. Liz is installing a tile backsplash. If each tile is a right isosceles triangle with two sides that measure 6 centimeters in length, how many tiles does she need to cover an area of 1800 square centimeters?

   A. 36 tiles

   B. 100 tiles

   C. 50 tiles

   D. 300 tiles

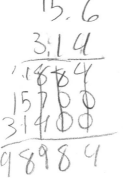

$\frac{1800}{18} = \frac{900}{9} = \frac{100}{1}$

9. If $\angle A$ measures $57°$, find $\angle G$.

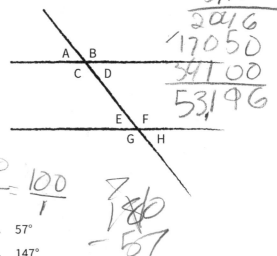

   A. $57°$

   B. $147°$

   C. $123°$

   D. $33°$

180
- 57
123

5. $2.31 \times 10^2 =$

   A. 23.1

   B. 231

   C. 2310

   D. 23100

15.6^2

3.14
15600
31400
98984

156²
3.41
2016
17050
34100
53196

10. What is 15% of 986?

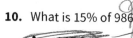

A. ~~146.9~~

B. ~~98.6~~

C. ~~9.86~~

D. 147.9

(handwritten: $986 \times 15 = $ ... $9863 \over 15$, 90, 1200, 13500, 14790)

11. 50% of 94 is:

A. 42

B. 52

C. 45

D. 47

(handwritten: 47)

12. The table below shows the number of hours worked by employees during the week. What is the median number of hours worked per week by the employees?

EMPLOYEE	HOURS WORKED
Suzanne	~~42~~
Joe	~~38~~
Mark	~~25~~
Ellen	50
Jill	~~45~~
Rob	~~46~~
Nicole	~~17~~
Dina	41

(handwritten: 17, 25, 38, 41, 42, 45, 46, 50)

A. 38

B. 41

C. 42

D. 41.5

13. Multiply the following terms: $(11xy)(2x^2y)$

A. $13xy + x$

B. $22x^3y^2$

C. $44x^3y^3$

D. $22xy^2 + 2x^2$

(handwritten: $22x^3y^2$)

14. $y = 2x - 5$, $x = 10$. What is y?

A. 10

B. 20

C. 15

D. 5

(handwritten: $y = 2(10) - 5$, $20 - 5$)

15. $x = 2$, $y = -3$, $z = 4$. Solve $x + y \times z$

A. -4

B. 10

C. -12

D. -10

(handwritten: $2 + {}^-3 \cdot 4$, $2 - 12$)

16. Factor the expression $64 - 100x^2$.

A. $(8 + 10x)(8 - 10x)$

B. $(8 + 10x)^2$

C. $(8 - 10x)^2$

D. $(8 + 10x)(8x + 10)$

17. Which expression would you solve first in the following: $(9 + 9) \times 987 + 4^6$

A. 4^6

B. $(9 + 9)$

C. ~~9×987~~

D. ~~$987 + 4$~~

18. Solve for y: $10y - 8 - 2y = 4y - 22 + 5y$

A. $y = -4\frac{2}{3}$

B. $y = 14$

C. $y = 30$

D. $y = -30$

(handwritten: $8y - 8 = 9y - 22$, $-8 = y - 22$, $14 = y$)

19. Solve for x: $(2x + 6)(3x - 15) = 0$

A. $x = -5, 3$

B. $x = -3, 5$

C. $x = -2, -3$

D. $x = -6, 15$

(handwritten: $6x^2$, $2x + 6 = 0$, $2x = -6$, $x = -3$, $3x = 15$, $x = 5$)

20. Round 0.1938562 to the nearest tenth.

A. 0.0

B. 0.2

C. 0.19

D. 0.194

21. Points <u>B and C</u> are on a circle, and a chord is formed by line segment $\overline{BC}$. If the distance from the center of the circle to point B is 10 centimeters, and the distance from the center of the circle to the center of line segment $\overline{BC}$ is 8 centimeters, what is the length of line segment $\overline{BC}$?

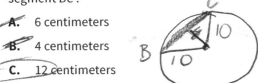

- **A.** 6 centimeters
- **B.** 4 centimeters
- **C.** 12 centimeters
- **D.** 14 centimeters

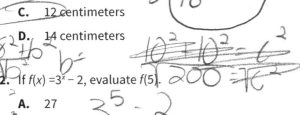

22. If $f(x) = 3^x - 2$, evaluate $f(5)$.

- **A.** 27
- **B.** 243
- **C.** 241
- **D.** 13

23. If a spherical water balloon is filled with 113 milliliters of water, what is the approximate radius of the balloon? (Note: The volume, V, of a sphere with radius r is found using the equation $V = \frac{4}{3}\pi r^3$.)

- **A.** 4.0 centimeters
- **B.** 3.0 centimeters
- **C.** 3.6 centimeters
- **D.** 3.3 centimeters

24. Simplify $\frac{13}{26}$ into a decimal.

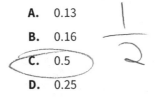

- **A.** 0.13
- **B.** 0.16
- **C.** 0.5
- **D.** 0.25

25. Factor the expression $100x^2 + 25x$ using a greatest common factor.

- **A.** $100x(x + 25x)$
- **B.** $25(4x + x)$
- **C.** $25x(4x + 1)$
- **D.** $25(4x^2 + x)$

GO ON

Electronics Information

1. What does the following image represent?

   A. voltage meter
   B. RAM
   C. computer
   D. integrated circuit

2. An ohm is used to measure what?

   A. force
   B. resistance
   C. power
   D. voltage

3. Which of the following devices would be able to turn electrical energy into sound waves?

   A. speaker
   B. wires
   C. tuning fork
   D. fire

4. What would a squiggly line in a circuit diagram represent?

   A. resistor
   B. conductor
   C. wire
   D. ground line

5. Which of the following is typically used to measure electrical energy?

   A. current
   B. volts
   C. kilowatt-hours
   D. watt-ohms

6. The _____ is the part of a given circuit which carries no voltage. Fill in the blank.

   A. fuse
   B. wire
   C. load
   D. ground

7. What force can be created by a current of electricity moving through a wire?

   A. magnetic force
   B. friction
   C. gravity
   D. resistance

8. What will be the most common result of a short circuit?

   A. light bulbs will explode
   B. a circuit breaker will trip
   C. the wire will cool down very quickly
   D. none of the above

9. The word *circuit* is commonly used in electronics. What is it?

   A. electricity
   B. a computer chip
   C. the path a current follows
   D. how voltage moves

10. The movement of which subatomic particle is responsible for electricity?

   A. protons
   B. electrons
   C. neutrons
   D. nuclei

11. When electrons move between two points, what is being created?

   A. pressure
   B. voltmeter
   C. force
   D. current

12. If the electrical potential is high at one end of a wire and low at the other, what will happen?

   A. current will flow to the low end
   B. current will flow to the high end
   C. the current will escape
   D. pressure will build, causing a fire

13. Changing AC to DC requires the use of a…

   A. transformer
   B. current switcher
   C. current manufacturing plant
   D. breaker box

14. Which one of these would be the least effective as a conductor?

   A. lead
   B. wood
   C. water
   D. copper

15. When a current goes over a certain value, a _____ can be installed which will melt and break the circuit.

   A. wire breaker
   B. breaker
   C. fuse
   D. current stopper

16. What is the definition of voltage?

   A. electricity moving
   B. electric tension
   C. electric kinetic energy
   D. electric pressure

17. Which of the following is used to measure voltage?

   A. voltage test unit
   B. voltage regulator
   C. voltmeter
   D. voltmeter

18. One joule per coulomb is also known as:

   A. 1 volt
   B. 1 ohm
   C. 1 resistor
   D. 1 potential

19. What makes integrated circuits possible?

   A. semiconductors
   B. metal
   C. conductors
   D. gold leaf

20. What common household device might you find a filament?

   A. integrated circuit
   B. microwave
   C. lamp
   D. computer

Auto and Shop Information

1. Four stroke engine cycles work in specific ways. They follow the intake - _____ - power – exhaust cycles. What is the part of the cycle that belongs in the blank space in this four-stroke engine cycle?

   A. compression
   B. decompression
   C. combustion
   D. power intake

2. If you were going to use a bolt to fasten two things together, what would you have to use to tighten the connection?

   A. screw
   B. drill bit
   C. ratchet
   D. nut

3. What is one of the possible reasons that a vehicle might have a better fuel efficiency rating?

   A. The gas costs less.
   B. The gas tank is larger.
   C. The vehicle costs less.
   D. The vehicle is lighter.

4. What would you use a plane for?

   A. to remove large portions of material
   B. to remove small bits of material, smoothing
   C. to roughen a surface
   D. to make things fit tighter

5. If someone told you that their vehicle had an inline 6 for the engine, what would they be indicating?

   A. The vehicle is lined up from all 6 directions.
   B. There are 6 cylinders.
   C. The engine is 6 feet long, in a single line.
   D. The gas tank is in line with the vehicle and holds 6 gallons.

6. What voltage would you expect to find in a typical commercial car battery sold in your local auto parts store?

   A. 6 volts
   B. 14 volts
   C. 12 volts
   D. 8 volts

7. If you wanted to measure how tight a bolt or a nut was, which one of the following tools would be the best choice for that?

   A. torque wrench
   B. screwdriver
   C. tight checker
   D. tape measure

8. What is a hand drill best used for?

   A. driving nails
   B. making holes in metal
   C. making holes in wood
   D. fastening tools

9. What might you use to tighten the object shown in the image below?

   A. hammer
   B. screwdriver
   C. metal drill
   D. drill press

10. Which part of the engine opens up in order to release the exhaust that has been created by combustion inside the cylinder?

    A. piston ring

    B. exhaust pipe

    C. intake valve

    D. exhaust valve

11. What is one of the components that you might find inside of a car battery that helps to produce the electricity the vehicle uses?

    A. muriatic acid

    B. sulfuric acid

    C. hydrochloric acid

    D. heavy peroxide

12. Emissions are the method through which vehicles get rid of their exhaust into the atmosphere. This can be potentially damaging to the environment as a whole. Which of the following is meant to help control emissions in vehicles?

    A. catalytic converter

    B. pump

    C. flywheel

    D. windshield

13. When you buy motor oil for your engine, there is frequently a W in the type of viscosity the oil is rated for. What does that W stand for? Example: 5W-20.

    A. windy

    B. wind chill

    C. winter

    D. water resistance

14. What type of fastener would you commonly use to hold two pieces of wood together for a quick and dirty project?

    A. nails

    B. screws

    C. bolts

    D. glue

15. How is the gasoline that reaches the cylinders of the engine controlled?

    A. ignition

    B. fuel tank

    C. throttle

    D. brake system

16. The steering wheel uses the _____ to turn the wheels.

    A. engine

    B. tie rod

    C. transmission

    D. brake system

17. The image below represents an important part of vehicles. What part is it?

    A. engine

    B. undercarriage

    C. fuel system

    D. transmission

18. Which of the following systems applies friction to the wheels of a vehicle?

    A. brake system

    B. transmission

    C. drums and pads

    D. padding system

19. What is used to create the combustion inside of an engine?

    A. fuel tank

    B. gasoline and air

    C. gas pedal

    D. oil and air

20. Force applied to an object causing it to twist is called what:

    A. twist force

    B. friction

    C. turning power

    D. torque

21. What needs to be added to the radiator of an engine?

    A. water and coolant

    B. water

    C. water and ice

    D. radiator replacement fluid

22. What is shown below?

    A. staples

    B. hammers

    C. nails

    D. screws

23. Which of the following is used in the engine to prevent metal parts from rubbing against one another and causing damage?

    A. rubber

    B. oil

    C. coolant

    D. air

24. What is one of the possible uses for a wrench?

    A. loosen a nut

    B. tighten a screw

    C. break metal down into smaller parts

    D. fix a piece of wood

25. What helps you guide the way that the material goes through a table saw?

    A. fence

    B. gutter

    C. saw blocker

    D. both hands at the same time

Mechanical Comprehension

1. What is the formula that is used to calculate work?

    A. $W = F \times s$

    B. $W = v \times F$

    C. $W = P \times s$

    D. $W = P \times F$

2. If an engine with a power output of around 2 horsepower is 95% efficient, what would the actual power output be, in horsepower?

    A. 190

    B. 95

    C. 1.90

    D. 0.19

3. A class two lever has the load placed between the fulcrum/pivot point and the effort being placed on it. Which of these might be an example of this?

    A. wheelbarrow

    B. gun

    C. wrench

    D. screwdriver

4. One of the following materials is a ceramic, which one is it?

    A. dirt

    B. gold

    C. pots

    D. brick

5. Force per unit of distance is a description of what?

    A. velocity

    B. force fields

    C. power

    D. work

6. Which one of the following might be a good example of a simple machine?

    A. ladder

    B. drill

    C. jackhammer

    D. iPod

7. A machine is operating with an input (for work) of 215-foot pounds. The output of the work for this machine is 204.25-foot pounds. What efficiency does this machine have, considering the information above?

    A. 90%

    B. 95%

    C. 100%

    D. 200%

8. If there are 20 lbs. on one side of a fulcrum (with equal lengths on both sides), which of the following combinations of weights would be enough to balance the loads on that fulcrum?

    A. 18 and 1

    B. 18 and 2

    C. 18 and 3

    D. 12 and 18

9. What kind of machine would a cam be considered?

    A. difficult

    B. simple

    C. conductor

    D. compound

10. How would you find an exerted force?

    A. find the force using the formula for work

    B. use the force field formula

    C. multiply applied force by the ratio of the areas to which it is being applied

    D. none of the above

11. What is a linkage?

    A. a way of converting rotating motion of a crank

    B. a type of fence

    C. a way to move chains

    D. a pulley system of complex design

12. Which of the following is a description of mechanical advantage?

    A. input force × output force

    B. output force × input force

    C. $\frac{\text{output force}}{\text{input force}}$

    D. $\frac{\text{input force}}{\text{output force}}$

13. If someone puts in 50 newtons of force and gets back 250 newtons of force, then what is the mechanical advantage?

    A. 5

    B. 10

    C. 15

    D. 20

14. What is the name of the force between objects that attracts them together?

    A. friction

    B. gravity

    C. force

    D. power

15. What might cause an object to accelerate?

    A. force

    B. gravity

    C. pulling on it

    D. all of the above

16. What is heat?

    A. a type of motion

    B. a type of pressure

    C. a type of energy

    D. the result of friction

17. What amount of force would have to be applied to move a box 25 meters? 55,000 joules worth of work is utilized in the process of moving the box.

    A. 2500 newtons

    B. 4000 newtons

    C. 2200 newtons

    D. 5500 newtons

18. If two liquids that have different densities are mixed together, what will happen?

    A. they will separate

    B. they will combine into one fluid

    C. they will react violently

    D. they will flow out of the container

19. What is the SI unit that is commonly used to measure mass?

    A. liter

    B. kilograms

    C. newtons

    D. all of the above

20. When would a spring likely be utilized?

    A. when making a large volleyball

    B. when creating a football

    C. when making a new baseball bat

    D. when building a pogo stick

21. When people are using a seesaw, the seesaw will work most efficiently if the two people have the same weight. Why?

    A. principle of equilibrium

    B. principle of force

    C. Newton's law

    D. the first law of motion

22. How do brakes slow vehicles down?

    A. force

    B. combustion

    C. acceleration

    D. friction

23. Shock absorption on vehicles is attributed to what?

    A. elasticity of springs

    B. brakes

    C. the engine block

    D. the weight of the vehicle

24. What type of device would you compare a crane to?

    A. car

    B. elevator

    C. pulley

    D. lever

25. If you were going to make something that was solid but would not float, what might you use?

    A. plastic

    B. glass

    C. metal

    D. wood

GO ON

Assembling Objects

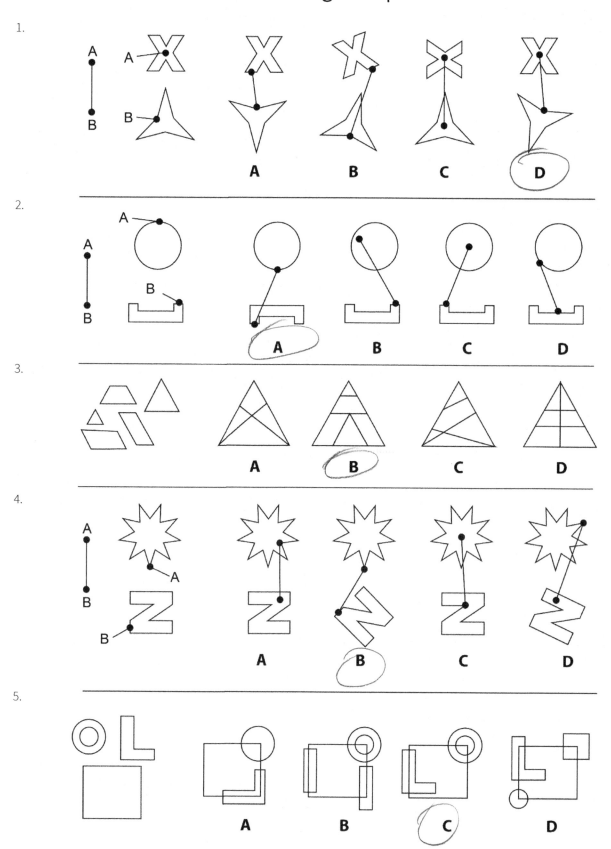

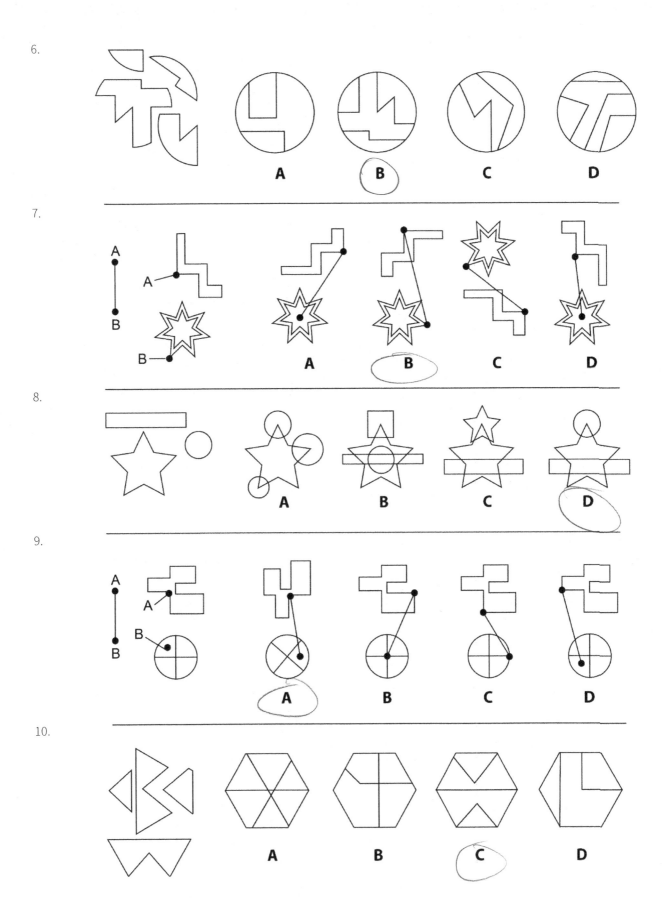

6.

A   B   C   D

7.

A   B   C   D

8.

A   B   C   D

9.

A   B   C   D

10.

A   B   C   D

11.

    A   A   B

A   **B**   **C**   **D**

12.

A   **B**   **C**   **D**

13.

A   **B**   **C**   **D**

14.

A   **B**   **C**   **D**

15.

A   **B**   **C**   **D**

16.

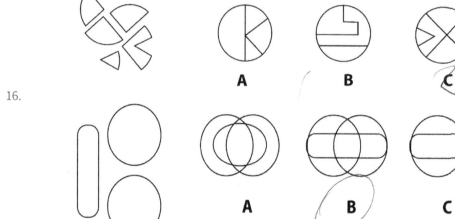

A   **B**   **C**   **D**

Practice Test One Answer Key

$\dfrac{-53}{216}$

GENERAL SCIENCE

1. C.	8. C.	15. C.	22. A.
2. A.	9. B.	16. A.	23. C.
3. D.	10. C.	17. D.	24. B.
4. B.	11. D.	18. C.	25. A.
5. A.	12. B.	19. D.	
6. B.	13. A.	20. D.	
7. A.	14. B.	21. B.	

75.5%

ARITHMETIC REASONING

1. B.	9. C.	17. B.	25. D.
2. C.	10. A.	18. A.	26. D.
3. D.	11. D.	19. C.	27. A.
4. B.	12. D.	20. C.	28. A.
5. A.	13. B.	21. B.	29. A.
6. D.	14. C.	22. D.	30. C.
7. B.	15. D.	23. A.	
8. B.	16. D.	24. A.	

WORD KNOWLEDGE

1. D.	10. C.	19. A.	28. B.
2. C.	11. A.	20. D.	29. D.
3. A.	12. D.	21. C.	30. B.
4. D.	13. B.	22. B.	31. C.
5. A.	14. D.	23. A.	32. D.
6. B.	15. C.	24. B.	33. C.
7. C.	16. A.	25. C.	34. B.
8. B.	17. B.	26. D.	35. B.
9. C.	18. B.	27. A.	

PARAGRAPH COMPREHENSION

1. C.	5. A.	9. D.	13. A.
2. B.	6. A.	10. A.	14. D.
3. D.	7. B.	11. B.	15. D.
4. C.	8. C.	12. D.	

Mathematics Knowledge

1.	A.	8.	A.	15.	D.	22.	C.
2.	B.	9.	C.	16.	A.	23.	B.
3.	A.	10.	D.	17.	B	24.	C.
4.	B.	11.	D.	18.	B.	25.	C.
5.	B.	12.	D.	19.	B.		
6.	B.	13.	B.	20.	B.		
7.	A.	14.	C.	21.	C.		

Electronics Information

1.	D.	6.	D.	11.	D.	16.	B.
2.	B.	7.	A.	12.	A.	17.	D.
3.	A.	8.	B.	13.	A.	18.	A.
4.	A.	9.	C.	14.	B.	19.	A.
5.	C.	10.	B.	15.	C.	20.	C.

Auto and Shop Information

1.	A.	8.	C.	15.	C.	22.	C.
2.	D.	9.	B.	16.	B.	23.	B.
3.	D.	10.	D.	17.	D.	24.	A.
4.	B.	11.	B.	18.	A.	25.	A.
5.	B.	12.	A.	19.	B.		
6.	C.	13.	C.	20.	D.		
7.	A.	14.	A.	21.	A.		

Mechanical Comprehension

1.	A.	8.	B.	15.	D.	22.	D.
2.	C.	9.	D.	16.	C.	23.	A.
3.	A.	10.	C.	17.	C.	24.	B.
4.	D.	11.	A.	18.	A.	25.	C.
5.	D.	12.	C.	19.	B.		
6.	A.	13.	A.	20.	D.		
7.	B.	14.	B.	21.	A.		

Assembling Objects

1.	D.	5.	C.	9.	A.	13.	D.
2.	A.	6.	B.	10.	C.	14.	B.
3.	B.	7.	B.	11.	C.	15.	C.
4.	B.	8.	D.	12.	A.	16.	B.

fourteen

PRACTICE TEST TWO

General Science

1. Which of the following is not in the Kingdom Plantae?

   A. cactus

   B. algae

   C. oak Tree

   D. sunflower

2. What is the primary difference between a cell membrane and a cell wall?

   A. A cell membrane is flexible, and a cell wall is rigid.

   B. A cell membrane is not found in plants, whereas a cell wall is.

   C. A cell membrane is not found in animals, whereas a cell wall is.

   D. A cell membrane is composed of protein, whereas a cell wall is composed of sugar.

3. Plants are autotrophs, meaning that they:

   A. consume organic material produced by animals

   B. produce their own food

   C. are able to move by themselves

   D. can automatically transform from a seed into a plant

4. Which of the following is not true of a virus?

   A. Viruses have DNA.

   B. Viruses do not have a nucleus.

   C. Viruses cannot survive without water.

   D. Viruses can be infectious.

5. In the digestive system, the majority of nutrients are absorbed in the:

   A. esophagus

   B. stomach

   C. small Intestine

   D. large Intestine

6. How many pairs of human chromosomes exist?

   A. 17

   B. 13

   C. 23

   D. 29

7. Animals engaging in a symbiotic relationship will do which of the following?

   A. help each other survive

   B. take one another's food

   C. attack one another

   D. eat each other

8. What organ system contains your skin?

    A. the respiratory system

    B. the epithelial system

    C. the lymphatic system

    D. the circulatory system

9. If a gene is expressed, then that means that:

    A. It is influencing a phenotype trait.

    B. It is being copied into another set of DNA.

    C. It will be passed on from mother to son.

    D. The gene will produce some hormones.

10. Which of the following structures is found in eukaryotes but not in prokaryotes?

    A. a cell wall

    B. mitochondria

    C. a nuclear membrane

    D. vacuoles

11. Upon touching a chair cushion and then a metal plate, John notices that the metal plate feels much colder than the cushion, although the surrounding air temperature is the same. What is an explanation for this?

    A. The chair cushion has a higher heat capacity than the metal plate.

    B. The metal plate has a higher heat transfer rate than the chair cushion.

    C. The metal plate is able to absorb more heat from the air than the cushion.

    D. The chair cushion produces some internal heat.

12. If a pitcher throws a baseball into the air and notices that it takes 5 seconds to reach its peak, how long will the baseball need to fall back to the ground? Neglect air resistance.

    A. 2.5 seconds

    B. 9.8 seconds

    C. 5.0 seconds

    D. 10.0 seconds

13. Which of the following is correct regarding an aqueous substance?

    A. It is soluble in water.

    B. It is very reactive.

    C. It is soluble in hydrocarbon.

    D. It is able to dissolve most other compounds.

14. In order for work to be performed, a force has to be:

    A. applied to an object

    B. applied to a surface

    C. applied to a moving object

    D. applied over a distance to an object.

15. During which of the following geologic periods did the majority of life develop?

    A. Triassic

    B. Permian

    C. Cretaceous

    D. Cambrian

16. The nearest star to the sun is about 4.2 light-years away and is known as:

    A. Alpha Centauri

    B. Barnard's Star

    C. Sirius A

    D. Proxima Centauri

Arithmetic Reasoning

1. Which of the following inequalities is true?

   A. 0.123 > 0.234

   B. −0.15 < −0.26

   C. −0.58 > 0.876

   D. −0.13 > −0.293

2. If you take 25% of 20, what would the resulting number be?

   A. 20

   B. 25

   C. 15

   D. 5

3. The average height of female students in a class is 64.5 inches, and the average height of male students in the class is 69 inches. If there are 1.5 times as many female students as male students, what is the average height for the entire class?

   A. 67.2 inches

   B. 66.75 inches

   C. 67.5 inches

   D. 66.3 inches

4. Meg rolled a 6-sided die 4 times, and her first 3 rolls were 1, 3, and 5. If the average of the 4 rolls is 2.5, what was the result of her fourth roll?

   A. 1

   B. 2

   C. 3

   D. 5

5. A parallelogram is divided into two triangles by drawing a straight line from one corner to an opposite corner. Which of the following is true of the two triangles?

   A. One triangle is an 180° rotation of the other.

   B. One triangle is a 90° rotation of the other.

   C. One triangle is an 180° reflection of the other.

   D. One triangle is a 90° translation of the other.

6. Jesse rides her bike 2 miles south and 8 miles east. She then takes the shortest possible route back home. What was the total distance she traveled?

   A. 17.75 miles

   B. 18.25 miles

   C. 8.25 miles

   D. 7.75 miles

7. Point A is x distance north of point B. Point C is east of point B and is twice as far from point B as point A is. What is the distance from point A to point C?

   A. 5x

   B. √3x

   C. 2x

   D. √5x

8. To get to school, Kaitlin walks 4 blocks north from her house, then turns right and walks 5 blocks east. How much shorter would her walk be if she could walk in a straight line from her house to her school?

   A. 6.4 blocks

   B. 3.2 blocks

   C. 6.0 blocks

   D. 2.6 blocks

9. A car rental company charges a daily fee of $48 plus 25% of the daily fee for every hour the car is late. If you rent a car for 2 days and bring it back 2 hours late, what will be the total charge?

A. $120

B. $108

C. $72

D. $144

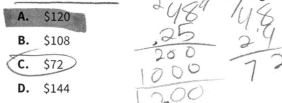

10. There are 450 students in the 10th grade; of these, 46% are boys. If 21% of the girls have already turned 16, how many girls in the 10th grade are 16?

A. 47

B. 94

C. 51

D. 10

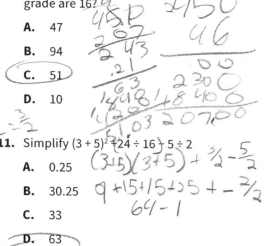

11. Simplify $(3 + 5)^2 + 24 \div 16 - 5 \div 2$

A. 0.25

B. 30.25

C. 33

D. 63

12. A marinade recipe calls for 2 tablespoons of lemon juice for every $\frac{1}{4}$ cup of olive oil. How much lemon juice would be used with $\frac{2}{3}$ cup olive oil?

A. $5\frac{1}{3}$ tablespoons

B. $\frac{3}{4}$ tablespoons

C. 4 tablespoons

D. $2\frac{1}{3}$ tablespoons

13. Put the following integers and fractions in order from smallest to largest: 0.125, $\frac{6}{9}$, $\frac{1}{7}$, 0.60

A. $\frac{1}{7}$, 0.125, $\frac{6}{9}$, 0.60

B. $\frac{1}{7}$, 0.125, 0.60, $\frac{6}{9}$

C. 0.125, $\frac{1}{7}$, 0.60, $\frac{6}{9}$

D. $\frac{1}{7}$, 0.125, $\frac{6}{9}$, 0.60

14. Megan cuts her birthday cake into 16 pieces. She and her 3 friends each eat a piece, and then Megan's dad eats $\frac{1}{3}$ of what is remaining. What fraction of the cake is left?

A. $\frac{13}{24}$

B. $\frac{1}{2}$

C. $\frac{1}{4}$

D. $\frac{3}{4}$

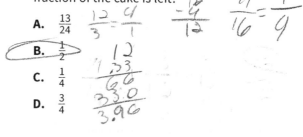

15. How many digits are in the sum 951.4 + 98.908 + 1.053?

A. 4

B. 5

C. 6

D. 7

16. Simplify: 0.08 × 0.12

A. 0.0096

B. 0.096

C. 0.96

D. 9.6

Word Knowledge

1. The soldiers engaged in <u>defensive</u> maneuvers to protect themselves from the enemy.
   - A. offensive
   - B. aggressive
   - C. protective *(circled)*
   - D. belligerent

2. <u>Tenacious</u> most nearly means:
   - A. surrendering
   - B. disloyal
   - C. persistent *(circled)*
   - D. thoughtful

3. <u>Complacent</u> most nearly means:
   - A. careless
   - B. careful *(circled)*
   - C. passionate
   - D. unconcerned *(highlighted)*

4. <u>Fidelity</u> most closely means:
   - A. falsity
   - B. treachery
   - C. separation
   - D. loyalty *(circled)*

5. The speech was <u>succinct</u>.
   - A. long
   - B. dull *(circled)*
   - C. exciting
   - D. short *(highlighted)*

6. <u>Venerate</u> most nearly means:
   - A. revere *(highlighted)*
   - B. admire *(circled)*
   - C. slander
   - D. desecrate

7. The <u>pivotal</u> moment in the battle changed the course of the war.
   - A. urgent
   - B. critical *(circled)*
   - C. minor
   - D. first

8. <u>Amicable</u> most closely means:
   - A. unfriendly
   - B. angry
   - C. friendly *(highlighted)*
   - D. honest *(circled)*

9. He <u>feigned</u> interest when questioned.
   - A. expressed
   - B. pretended *(circled)*
   - C. celebrated
   - D. raised

10. She <u>reigned</u> for twenty-seven years.
   - A. hid
   - B. worked
   - C. waited
   - D. ruled *(circled)*

11. <u>Coup</u> most nearly means:
   - A. takeover *(highlighted)*
   - B. henhouse *(circled)*
   - C. cover
   - D. failure

12. <u>Abate</u> most closely means:
   - A. increase *(circled)*
   - B. replace
   - C. decrease *(highlighted)*
   - D. worsen

13. Their efforts were in <u>vain</u>.

    A. conceited

    B. useless

    C. effective

    D. practical

14. The <u>yield</u> was better than expected.

    A. revenue

    B. surrender

    C. outcome

    D. product

15. <u>Compel</u> most closely means:

    A. force

    B. free

    C. encourage

    D. discourage

16. <u>Advocate</u> most nearly means:

    A. adversary

    B. enemy

    C. friend

    D. supporter

Paragraph Comprehension

1. Lieutenant Hiroo Onoda was a Japanese soldier who was sent to a small island in 1944 as an <u>emissary</u>. He refused to believe that Japan surrendered in WWII until his commanding officer finally traveled back to the island in 1974 and finally convinced him that the defeat was real. He then returned to Japan and received a hero's welcome. In this sentence what is the definition of emissary?

    A. Emissary refers to Hiroo Onoda being an ambassador for the Japanese army.

    B. In this sentence, emissary means a secret agent or spy.

    C. The word emissary means messenger in this sentence.

    D. Emissary, in the context of this sentence, means a delegate of the Japanese government meant to establish an embassy on the island.

2. Milton S. Hershey was the founder of North America's largest chocolate manufacturer, now known as, The Hershey Company. It is hard to believe that, with such a large, successful business, that Hershey's first attempts in the confectionary business were such failures. After finishing a confectionary apprenticeship, he opened his own candy shop in Philadelphia; 6 years later it went out of business. He then returned home after failing to manufacture candies in New York City and in 1903 construction of a chocolate plant began in his hometown which was later renamed Hershey, Pennsylvania. What is the main message of this passage?

    A. As an entrepreneur, if your first idea fails, do not give up, but move on to your next plan for success.

    B. One can only be successful in starting a flourishing business with the support of your hometown.

    C. It is more successful to manufacture chocolate than candy.

    D. If you start a worldwide profitable business in your hometown, they will rename the town in your honor.

3. Beware the leader who bands the drums of war in order to whip the citizenry into a patriotic fervor, for patriotism is indeed a double-edged sword. This quote of Caesar's is completely <u>anachronistic</u>; what does *anachronistic* mean in this context?

    A. This word means stolen in this sentence. This is a quote from another ruler from the time of Caesar, but not Caesar himself.

    B. Anachronistic means a quote that is pieced together from parts of speeches made by an individual. It is, therefore, a quote without any real meaning.

    C. In this sentence, the word anachronistic means that this is a true and accurate quote; not a paraphrase.

    D. The word anachronistic is defined as a quote that is not historically accurate in its context. At the time of Caesar; there were no drums of war, for example.

4. A stitch in time saves nine. This is a proverbial expression that has used for hundreds of years. What is this phrase referring to?

    A. This expression means that there is a "rip" of some sort in time and space and that only by repairing this rip will we save the world.

    B. When this phrase is used, the person means that by repairing a piece of clothing, you will save $9.00 on replacing the garment.

    C. This phrase refers to a broken relationship. If it is not repaired in time, it will take years (maybe even 9 years) to mend.

    D. The literal meaning of this expression means that if you stitch something up in time, you will save 9 stitches later. In other words, if you don't procrastinate, and repair something as soon as it is required, you won't have a bigger or worse job to fix at a later time.

5. In the Shakespearean play, *Julius Caesar*, a soothsayer calls out to Caesar with the following quote; "Beware the Ides of March!" What did this declaration of the soothsayer mean?

   A. The soothsayer was warning the ruler of his impending betrayal and death at the hands of some of his most trusted men.

   B. This phrase was actually warning the crowd, not Caesar that on ever Ides of March the ruler must choose one human sacrifice to offer up to the Roman gods to guarantee prosperity for the coming year.

   C. The Ides of March was a day of celebration in the Roman Empire to commemorate the deaths of the Christians in the Coliseum. The soothsayer was merely thanking Caesar for the day of celebration. The word "Beware" has been shown to be translated incorrectly into English.

   D. The soothsayer meant to warn Caesar not to upset or anger the god for whom the month of March was named; Mars, the god of war. To upset the god Mars, was to ensure plague, famine, or other ruin.

6. The Schneider Family was not your average family. Three generations lived in one house; Mom and Dad, four of their children, and Mom's parents who were well into their golden years. The term *golden years* is a nice way of meaning what?

   A. The term *golden years* refers to the best years of someone's life.

   B. This phrase means that the mom's parents were old or elderly people.

   C. *Golden years* is another way of saying, when they were rich.

   D. In this paragraph, the meaning of the term *golden years* is that the grandparents were spending their years taking care of everyone else in the family.

7. Examples of colloquialisms include Facebook, y'all, gotta, and shoulda. What is the definition of a *colloquialism*?

   A. Words that are only used by Americans who live in the south.

   B. Words that only uneducated people say.

   C. Words that are used in an informal conversation, not a more formal discussion.

   D. Words that have recently been added to the dictionary as acceptable words to use in the American English Language.

8. Tornados occur when air begins to rotate and comes into contact with both the earth and a cloud at the same time. Although the size and shape of tornados vary widely, one can usually see a funnel stretching from the sky down to land. Most tornados are accompanied with winds as fast as 110 miles per hour and extreme ones can have winds as fast as 300 miles per hour. The path of a tornado is hard to predict, but it is becoming possible to detect them just before or as they form with the continued collection of data through radar and "storm chasers". Storm chasing is a dangerous profession, so why do people continue to put their lives in danger this way?

   A. Storm chasers are an interesting breed of people who seek the thrill and adventure that comes along with this profession, much like extreme sports.

   B. News channels will pay large sums of money for good video of tornados, so, although it is a dangerous profession, the money is worth the risk.

   C. It is very important to discover as much as possible about how tornados work so that ultimately, scientists will detect them earlier and give people more advanced warning to get to safety. More advanced warning is the only way more lives will be saved.

   D. For statistics reasons, it is important to get first-hand data during a tornado. This way they can be compared to other natural disasters such as hurricanes and tsunamis.

9. Of the phrases below, which one is an example of an oxymoron?

   A. Three of the employees were let go due to suspicion of stealing money from the cash drawer.

   B. The stormy night was perfect for this woman's current mood.

   C. It was raining cats and dogs when the school bell rang.

   D. The community center was collecting useless treasures for their upcoming garage sale.

10. Secret Santa Sings Special Song for Sweetheart is an example of alliteration. What does *alliteration* mean?

    A. Alliteration means that the sentence has more than one meaning.

    B. Alliteration means that people with a stutter would have difficulty saying this sentence.

    C. Alliteration means that most of the words in the sentence begin with the same letter.

    D. In this sentence *alliteration* means that a secret Santa literally sang a special song for his sweetheart; it means that this even actually happened.

11. Jim had been on the road for 36 hours straight to meet an important client and hopefully finalize a huge new account for his advertising agency. After checking into his hotel, he intended just to drop off his suitcases and go down to the restaurant for a late supper. Once he entered the room, however, the cozy couch looked so friendly and welcoming to the weary traveler. Personification is a literary device that gives human characteristics to a non-human object. What phrase in this paragraph is an example of personification?

    A. An example from this paragraph that is personification is, *the cozy couch looked so friendly and welcoming…*

    B. *Jim had been on the road for 36 hours straight…* is an example of personification in this paragraph.

    C. The phrase, *…and hopefully, finalize a huge new account for his advertising agency* is an example of personification.

    D. An example of personification, in this paragraph, is, *…to just drop off his suitcase and go down to the restaurant…*

GO ON

Mathematics Knowledge

1. $F(x) = 6x - 3$, $G(x) = 3x + 4$

    What will be $F(3) - G(2)$ equal to?

    A.   4
    B.   3
    C.   5
    D.   2

    $(6(3) - 3) - (3(2) + 4)$

    $(15) - (10)$

2. The mean of the marks obtained by the students in a class is 60 out of 100, and the standard deviation is 0. It means that

    A.   Half of the students have scored marks less than 60.

    B.   Half of the students have scored marks greater than 60.

    C.   No student has scored 100 marks.

    D.   All the students have scored 60 marks each.

3. 0.00092×10^{-3} is equal to which of the following?

    A. 0.000093×10^{-4}

    B. 0.000092×10^{-2}

    C. 0.000000092

    D. 0.92×10^{-8}

4. The remainder is 3 when we divide one number by another number. What can be these two numbers from the following?

    A.   9, 5
    B.   8, 5
    C.   9, 6
    D.   both B and C

5. If A and B are odd integers. Which of the following expressions must give an odd integer?

    A. $A \times B$
    B. $A + B$
    C. $A - B$
    D.   both A and C

6. $\frac{4}{5} \div \underline{\quad} = 2$

    Which of the following will fill the blank?

    A. $\frac{2}{5}$
    B. $\frac{5}{2}$
    C. $\frac{1}{5}$
    D.   both A and C

    $\frac{4}{5} \cdot \frac{5}{2} = \frac{20}{10}$

7. Given is a set {2, 4, 6, 8... 50}

    How many numbers in the given set are completely divisible by 3?

    A.   6
    B.   8
    C.   7
    D.   9

8. What will be the area of the shaded region in the given figure?

    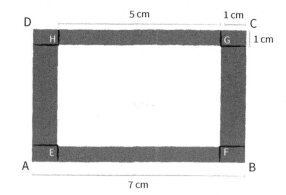

    A.   24 cm²
    B.   26 cm²
    C.   23 cm²
    D.   28 cm²

9. If $2x - y + 6 = 2$ then what will be the value of $6x$?

    A. $3y + 12$
    B. $y - 12$
    C. $y + 12$
    D. $3y - 12$

    $2x - y + 6 = 2$

    $2x - y = -4$

    $2x = -4 + y$

    $6x = -2 + \frac{y}{2} \cdot \frac{6}{1}$

    $-2 + 3y$

10. A point is located in coordinate system at $(1, 2)$. What will be the location of this point if it is shifted 5 units downwards and 3 units in the right direction?

A. $(6, -1)$

B. $(-4, 5)$

C. remains the same

D. $(4, -3)$

(handwritten: $(4,$

11. $\dfrac{y + 2}{3y^2 + 2y} + \dfrac{2y - 1}{6y^3 + 4y^2} =$

A. $\dfrac{2y^2 + 6y - 1}{6y^3 + 4y^2}$

B. $\dfrac{2y^2 + 8y - 1}{6y^3 + 4y^2}$

C. $\dfrac{2y^2 + 6y - 1}{3y^2 + 2y}$

D. $\dfrac{2y^2 + 8y - 1}{3y^2 + 2y}$

(handwritten work: $\dfrac{2y^2 + 4y + 2y - 1}{2y^2(3y+2)}\quad 2y^2(3y+2)$ $2y^2 + 6y - 1$)

12. If each side of the square has been increased by 1 cm and the area has now become 36 cm². What will be the length of one side of the square before?

A. 4 cm

B. 5 cm

C. 6 cm

D. 7 cm

13. $(9)^{-3} =$

A. $\dfrac{1}{9}$

B. $-\dfrac{1}{(9)^3}$

C. $\dfrac{1}{(9)^{-3}}$

D. $\dfrac{1}{(9)^3}$

14. What is the degree of polynomial $5x^2y - 5x^2y^2 + 5x^3y^2$?

A. 12

B. 4

C. 8

D. 5

15. Which one of the following numbers is not divisible by 3?

A. 2352

B. 3243

C. 6143

D. 5232

(handwritten long division work)

16. $(3 - x)(3 + x) =$

A. $9 - x^2$

B. $x^2 - 9$

C. $9 + x^2$

D. $x^2 - 6x + 9$

(handwritten: $9 + 3x - 3x - x^2$)

GO ON

Electronics Information

1. Which one of the following is the correct relation between Power (P), Voltage (V) and Current (I)?

   A. $P = \frac{V}{I}$

   B. $V = \frac{I}{P}$

   C. $P = VI$

   D. $P = \frac{I}{V}$

2. What is the role of a rectifier?

   A. It converts AC to DC.

   B. It steps up AC.

   C. It converts DC to AC.

   D. It steps up DC.

3. Lagging power factor in an electrical circuit occurs due to the presence of which of the following?

   A. capacitor

   B. inductor

   C. resistor

   D. transistor

4. RL-circuit denotes which one of the following?

   A. Resistor-Capacitor circuit

   B. Response-Lagging circuit

   C. Reverse-Lagging circuit

   D. Resistor-Inductor circuit

5. The reverse recovery of a typical MOSFET is:

   A. comparable to that of a BJT

   B. faster than that of a BJT

   C. slower than that of a BJT

   D. does not exist

6. A BJT, being used for amplification purpose in some applications, has:

   A. CB junction is reverse biased, and EB junction is forward biased.

   B. CB junction is forward biased, and EB junction is also forward biased.

   C. CB junction is reverse biased, and EB junction is also reverse biased.

   D. CB junction is forward biased, and EB is reverse biased.

7. Which of the following can be the applications of a transistor?

   A. switching device

   B. variable Resistor

   C. amplifier

   D. all of the above

8. The DC current gain of the transistor (i.e. β) is represented as:

   A. $\frac{I_B}{I_C}$

   B. $I_B \times I_C$

   C. $\frac{I_C}{I_B}$

   D. $I_C - I_B$

9. The collector current in a transistor is regulated by which of the following?

   A. base current

   B. emitter current

   C. input resistance

   D. output resistance

10. Which one of the following statements is correct?

A. The electrical conductivity of an intrinsic semiconductor is high as compared to that of an extrinsic semiconductor.

B. The electrical conductivity of an intrinsic semiconductor is low as compared to that of an extrinsic semiconductor.

C. The electrical conductivity of an intrinsic semiconductor is equal to that of an extrinsic semiconductor.

D. We cannot find a relationship between the electrical conductivity of intrinsic and extrinsic semiconductors.

11. Which one of the following describes the correct relationship between emitter current (I_E), Collector Current (I_C) and Base Current (I_B) of a transistor?

A. $I_E = I_C - I_B$

B. $I_E = \dfrac{I_B}{I_C}$

C. $I_E = I_C + I_B$

D. $I_E = \dfrac{I_C}{I_B}$

12. Two resistors of different values are connected in parallel to the supply voltage. Which of the following is the correct statement regarding this circuit?

A. The resistance having a large value causes more power loss.

B. The resistance having a small value causes more power loss.

C. Both resistors cause same power loss.

D. We cannot calculate power loss with the given data.

13. Which one of the following is the condition for an RLC circuit to be at resonance?

A. $XL = XC$

B. $XL = R$

C. $XC = R$

D. $XL > XC$

14. According to Ohm's Law, what will happen to the voltage if the resistance increases by 4 times and the current becomes half of its actual value?

A. The voltage becomes half of its actual value.

B. The voltage remains same

C. The voltage becomes twice its actual value.

D. The voltage becomes four times its actual value

15. Which one of the following statements is true regarding the characteristics of an ideal operational amplifier?

A. An ideal operational amplifier has an infinite output impedance.

B. An ideal operational amplifier has an infinite input impedance.

C. An ideal operational amplifier has a finite bandwidth.

D. all of the above

16. Which one of the following formulas depicts the output voltage of an inverting amplifier (where R_1 is the input resistance and R_2 is the output resistance)?

A. $V_{out} = V_{in} \times \left(\dfrac{R_2}{R_1} \right)$

B. $V_{out} = -V_{in} \times \left(\dfrac{R_1}{R_2} \right)$

C. $V_{out} = V_{in} \times \left(\dfrac{R_1}{R_2} \right)$

D. $V_{out} = -V_{in} \times \left(\dfrac{R_2}{R_1} \right)$

GO ON

Auto and Shop Information

1. A fuel-injected engine does not have:

    A. a fuel pump

    B. an intake valve

    C. a carburetor

    D. either a or c

2. What device steps up the voltage delivered to the distributor?

    A. contact points

    B. battery

    C. spark plug

    D. ignition coil

3. Which measurement can cause abnormal tire wear if not properly adjusted?

    A. toe

    B. camber

    C. tire pressure

    D. all of the above

4. Which of the following oils would be the thinnest at high temperatures?

    A. SAE-30

    B. 5W-20

    C. 5W-30

    D. 10W-40

5. Which of the following lists the stages of a four stroke engine cycle in the correct order?

    A. compression, intake, exhaust, power

    B. suction, compression, power, output

    C. intake, compression, power, exhaust

    D. intake, displacement, ignition, power

6. In which of the following engine configurations are the cylinders arranged in a single row?

    A. V

    B. inline

    C. flat

    D. both a and c

7. What does antifreeze do?

    A. raises the boiling point of water

    B. lowers the boiling point of water

    C. lowers the freezing point of water

    D. both a and c

8. In which type of vehicle construction is the body not permanently mounted to the frame?

    A. unibody

    B. body-over-frame

    C. monocoque

    D. all of the above

9. Which of these is not true of the power stroke?

    A. All valves are closed during the power stroke.

    B. The power stroke occurs before the compression stroke.

    C. Expanding gasses force the piston downward during the power stroke.

    D. The spark plug ignites the air-fuel mixture to initiate the power stroke.

10. What translates the linear motion of the piston to the rotation of the crankshaft in an engine?

    A. torque converter

    B. connecting rod

    C. catalytic converter

    D. push rod

11. What does the internal combustion engine burn?

    A. air

    B. fuel

    C. air-fuel mixture

    D. alcohol

12. What would be used to create external, or male, threads?

    A. tap

    B. die

    C. thread gauge

    D. calipers

13. Which of the following should not be used for tight bolts?

    A. box end wrench

    B. crescent wrench

    C. breaker bar

    D. six-point socket

14. Vernier calipers can be used to measure:

    A. diameter

    B. depth

    C. thickness

    D. all of the above

15. Which of the following should be used to begin the removal of a spring pin?

    A. drift punch

    B. center punch

    C. chisel

    D. none of the above

16. When arc welding, what is attached to the workpiece?

    A. electrode

    B. flux

    C. ground clamp

    D. filler rod

17. Which of the following wooden objects would be created on a lathe?

    A. table top

    B. frame

    C. baseball bat

    D. shelf

18. What is the purpose of a sanding block?

    A. to secure the workpiece while sanding

    B. to sand within small crevices of a wooden object

    C. to smooth the other surface of round objects

    D. to prevent an uneven finish when sanding

19. The mallet and sledge are examples of:

    A. cutting tools

    B. striking tools

    C. splitting tools

    D. flattening tools

20. What should be done when using a file?

    A. Apply pressure only on the forward stroke.

    B. Secure a handle onto the pointed tang.

    C. Hold the file with one hand on the handle and the other on the file tip.

    D. all of the above

21. What tool would be best suited for holding a small electrical component in a small, confined space?

    A. slip-joint pliers

    B. vise-grips

    C. needle-nose pliers

    D. calipers

22. Which of following can be used to cut metal without the application of heat?

    A. cold chisel

    B. pin punch

    C. hot chisel

    D. starting punch

Mechanical Comprehension

1. Which of the following is not a correct unit for the amount of work done?

   A. joule

   B. horsepower-hour

   C. calorie

   D. newton

2. Observe the figure:

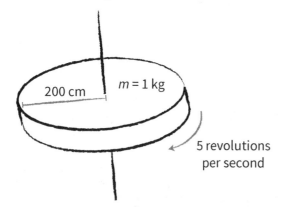

200 cm $m = 1$ kg

5 revolutions per second

The kinetic energy of the disc is:

   A. $80\,\pi^2$ J

   B. $100\,\pi^2$ J

   C. $125\,\pi^2$ J

   D. $144\,\pi^2$ J

4. In the following figure, consider a block of mass m. What is the ratio of the force required, for a person to lift the block upwards with and without a pulley? (Hint: Assume F = T)

   A. 2

   B. $\frac{1}{3}$

   C. 3

   D. $\frac{1}{2}$

8. The factor which distinguishes between a scalar and a vector quantity is:

   A. magnitude

   B. direction

   C. both A and B

   D. neither A nor B

3. Consider the following figure of a rolling wheel on smooth horizontal surface:

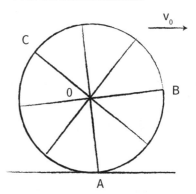

Then,

> **i.** Speed at the point A is 0
>
> **ii.** Speed at point B and C = v_o
>
> **iii.** Speed at point B > Speed at point O

   A. All the statements are true.

   B. Only statement (i) and (ii) are true.

   C. Only statement (i) and (iii) are true.

   D. Only statement (ii) and (iii) are true.

5. A block of mass 3kg lies on a horizontal surface with μ = 0.7, select the force closest to what is required just to move the block:

   A. 15N

   B. 21N

   C. 18N

   D. 24N

7. A ball is thrown into the air. After few seconds, it returns back to the earth. What can be its likely cause?

   A. Earth's gravitational field pulls it back.

   B. Its speed did not match the escape velocity of earth.

   C. Neither of these is correct.

   D. Both A and B are correct.

6. P is a block of mass 5kg. At point Q, a block of mass 3kg was attached just to slide the block P.

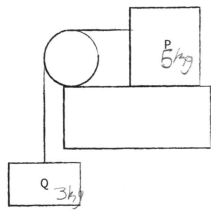

If no displacement occurs, the coefficient of friction between the block P and the horizontal surface is:

- **A.** 0.5
- **B.** 0.6
- **C.** 0.7
- **D.** 0.8

9. An athlete couldn't stop himself immediately after crossing the finish line. He was explained why this was happening by Newton's:

- **A.** 1st law of motion
- **B.** 2nd law of motion
- **C.** 3rd law of motion
- **D.** Law of Universal Gravitation

10. How is the weight of a person in an elevator affected if the elevator accelerates upwards, accelerates downwards and is at rest?

- **A.** increases, decreases, remains constant
- **B.** decreases, remains constant, increases
- **C.** remains constant, increases, decreasess
- **D.** decreases, increases, remains constant

12. The threads of a screw work on the principle of another type of simple machine, which is:

- **A.** lever
- **B.** inclined plane
- **C.** wedge
- **D.** none of the above

13. The shaft of the screw penetrates wood through the principle of yet another simple machine, which is:

- **A.** inclined plane
- **B.** lever
- **C.** wedge
- **D.** none of the above

14. Forceps, scissors, fishing rod, bottle opener

The above objects are an example of which order of the lever:

- **A.** 3rd, 2nd, 3rd, 1st
- **B.** 2nd, 3rd, 1st, 3rd
- **C.** 3rd, 1st, 3rd, 2nd
- **D.** 1st, 3rd, 2nd, 3rd

15. A mechanic observes that he is able to lift the car by 2cm if he moves the lever down by 30cm. if he is applying a force of 20N to the lever, the force applied by the lever on the car is:

- **A.** 250N
- **B.** 300N
- **C.** 350N
- **D.** 400N

16. Angular momentum of a body doesn't change if:

- **A.** External torque is not applied.
- **B.** External torque is applied in CW Direction.
- **C.** External torque is applied in CCW Direction.
- **D.** External torque has no effect on the angular momentum of the body.

GO ON

Assembling Objects

1.

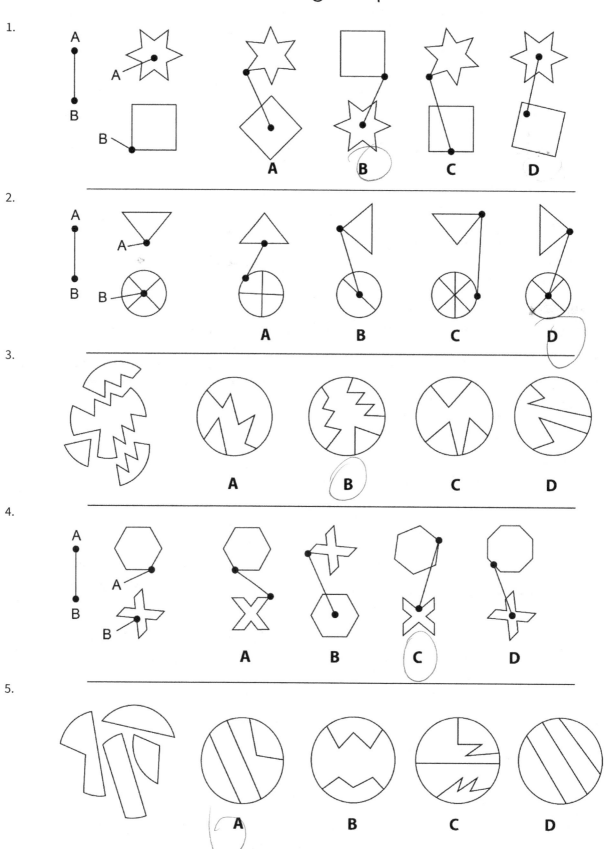

2.

3.

4.

5.

6.

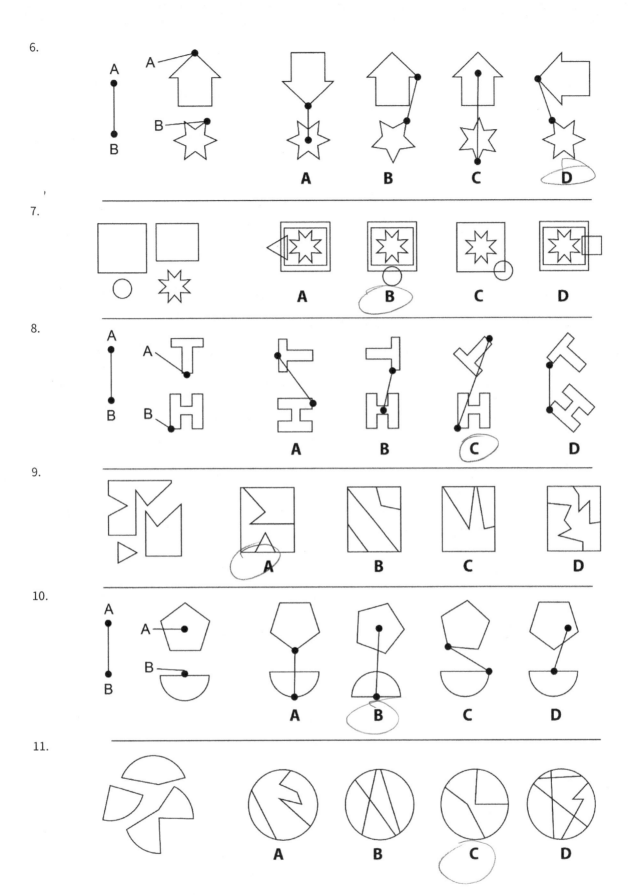

7.

8.

9.

10.

11.

12.

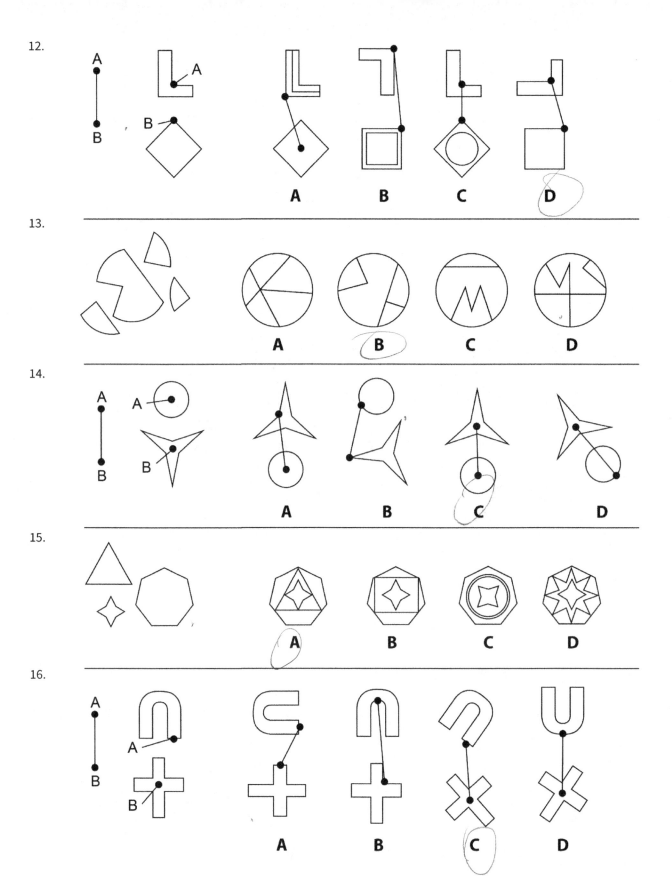

13.

14.

15.

16.

Practice Test Two Answer Key

GENERAL SCIENCE

1. B.	5. C.	9. A.	13. A.
2. A.	6. C.	10. C.	14. D.
3. B.	7. A.	11. B.	15. D.
4. C.	8. B.	12. C.	16. D.

ARITHMETIC REASONING

1. D.	5. A.	9. A.	13. C.
2. D.	6. B.	10. C.	14. B.
3. D.	7. D.	11. D.	15. D.
4. A.	8. D.	12. A.	16. A.

63.6%

WORD KNOWLEDGE

1. C.	5. D.	9. B.	13. B.
2. C.	6. A.	10. D.	14. A.
3. D.	7. B.	11. A.	15. A.
4. D.	8. C.	12. C.	16. D.

PARAGRAPH COMPREHENSION

1. B.	4. D.	7. C.	10. C.
2. A.	5. A.	8. C.	11. A.
3. D.	6. B.	9. D.	

MATHEMATICS KNOWLEDGE

1. C.	5. A.	9. D.	13. D.
2. D.	6. A.	10. D.	14. D.
3. B.	7. B.	11. A.	15. C.
4. D.	8. A.	12. B.	16. A.

ELECTRONICS INFORMATION

1. C.	5. B.	9. A.	13. A.
2. A	6. A.	10. B.	14. C.
3. B.	7. D.	11. C.	15. B.
4. D.	8. C.	12. B.	16. D.

Auto and Shop Information

1. C.	7. D.	13. B.	19. B.
2. D.	8. B.	14. D.	20. D.
3. D.	9. B.	15. A.	21. C.
4. B.	10. B.	16. C.	22. A.
5. C.	11. C.	17. C.	
6. B.	12. B.	18. D.	

Mechanical Comprehension

1. D.	5. C.	9. A.	13. C.
2. B.	6. B.	10. A.	14. B.
3. D.	7. D.	11. B.	15. A.
4. B.	8. B.	12. C.	

Assembling Objects

1. B.	5. A.	9. A.	13. B.
2. D.	6. D.	10. B.	14. C.
3. B.	7. B.	11. C.	15. A.
4. C.	8. C.	12. D.	16. C.

fifteen

PRACTICE TEST THREE

General Science

1. A mutation in DNA can be caused by all of the following except:

   A. ultraviolet radiation

   B. chemical exposure

   C. DNA replication error

   D. exonic duplication

2. People who suffer from Type I diabetes are lacking function in which organ?

   A. liver

   B. pancreas

   C. stomach

   D. heart

3. One of the primary differences between fungi and plants is that:

   A. Fungi can produce their own food and plants cannot.

   B. Plants have chlorophyll and fungi do not.

   C. Fungi are able to grow without water and plants cannot.

   D. Fungi and plants have no major differences.

4. Which of the following organisms is capable of asexual reproduction?

   A. squash plant

   B. amoeba

   C. salmon

   D. koala bear

5. In our atmosphere, nitrogen is the most common element, and makes up approximately what percentage?

   A. 25%

   B. 51%

   C. 65%

   D. 78%

6. In the human body, which of the following is responsible for clotting blood?

   A. platelets

   B. white blood cells

   C. red blood cells

   D. osteoplasts

7. In plants, the female reproductive structures reside in the pistil, whereas the male reproductive structures are in the:

   A. stamen
   B. anther
   C. sepals
   D. petals

8. In the human body, communication occurring from cell to cell can happen through the use of:

   A. neurotransmitters
   B. pili
   C. flagella
   D. ATP

9. In the following list, which would be considered to be at the top of the food chain?

   A. snake
   B. mouse
   C. hawk
   D. tomato plant

10. Tundra, rainforest, and prairie are all examples of ecological classifications known as:

    A. ecomes
    B. partitions
    C. biomes
    D. communities

11. According to electron theory, what is the maximum number of bonds a carbon atom can have?

    A. two
    B. three
    C. four
    D. five

12. If a rowboat weighs 50 kilograms, how much water needs to be displaced in order for the boat to float?

    A. 25 liters
    B. 50 liters
    C. 100 liters
    D. 500 liters

13. Given the reaction: $CaSO_{4(aq)} + CuCl_{2(aq)} \rightarrow$, which of the following is a possible product?

    A. CaCl
    B. CaCu
    C. $CuSO_3$
    D. $CaCl_2$

14. A ball with a mass of 0.5 kg is moving at 10 m/s. How much kinetic energy does it have?

    A. 15 Joules
    B. 25 Joules
    C. 50 Joules
    D. 55.5 Joules

15. A volcano that is low and flat to the ground, and does not typically have large, violent eruptions can be classified as what?

    A. a plane volcano
    B. a cinder cone
    C. a shield volcano
    D. a screen volcano

16. Which of these types of rocks is created near or on the earth's surface?

    A. igneous rock
    B. sedimentary rock
    C. crustaceous rock
    D. metamorphic rock

Arithmetic Reasoning

1. A car dealership is offering huge deals for the weekend. The commercials claim that this year's models are 20% off the list price, and the dealership will pay the first 3 monthly payments. If a car is listed for $26,580, and the monthly payments are set at $250, what are the total potential savings?

    A. $20,514

    B. $5,566

    C. $6,066

    D. $1,282

2. Joe baked brownies in a 9 inch × 11 inch × 2 inch tray. He then cut the brownies into 12 large pieces. Joe ate 2 pieces, his roommate ate 3 pieces, and the dog, unfortunately, ate half of what was remaining. How much of the brownies did the dog eat?

    A. 99 in³

    B. 57.75 in³

    C. 82.5 in³

    D. 115.5 in³

3. The county is instituting a new license plate system. The new plates will have 6 digits: the first digit will be 1, 2, or 3, and the next 5 digits can be any number from 0 – 9. How many possible unique combinations does this new system offer?

    A. $3^3 \times 10^5$

    B. 3×10^6

    C. 10^6

    D. 53

4. What percent of 14 is 35?

    A. 4.9%

    B. 2.5%

    C. 40%

    D. 250%

5. The high temperature on Wednesday is 4 degrees warmer than the high temperature on Tuesday, which was 5 degrees cooler than the high temperature on Monday. If the high temperature on Thursday is predicted to be 3 degrees cooler than the high on Wednesday, what is the difference in temperature between Monday and Thursday?

    A. 3 degrees

    B. 4 degrees

    C. 6 degrees

    D. 12 degrees

6. A dry cleaner charges $3 per shirt, $6 per pair of pants, and an extra $5 per item for mending. Annie drops off 5 shirts and 4 pairs of pants, 2 of which needed mending. Assuming the cleaner charges an 8% sales tax, what will be Annie's total bill?

    A. $52.92

    B. $49.00

    C. $45.08

    D. $88.20

7. A car dealership has sedans, SUVs, and minivans in a ratio of 6:3:1, respectively. In total, there are 200 of these vehicles on the lot. What proportion of those vehicles are sedans?

    A. 120

    B. $\frac{3}{100}$

    C. $\frac{3}{5}$

    D. $\frac{3}{10}$

8. A summer camp requires a ratio of 1 camp counselor to every 6 campers. Each camp counselor makes $480 per week. If the camp director wants to register an additional 30 campers for 2 weeks this summer, how much more will she have to budget to pay counselors?

    A. $2,400

    B. $4,800

    C. $5,760

    D. $2,880

9. Simplify: $(x^{\frac{1}{2}})^{-3}$

$X^{-\frac{5}{2}}$

    A. $x^{\frac{-1}{2}}$

    B. $x^{\frac{-5}{2}}$

    C. $\frac{1}{\sqrt{x^3}}$

    D. $\sqrt{x^3}$

10. A sporting goods store is offering an additional 30% off all clearance items. Angie purchases a pair of running shoes on clearance for $65.00. If the shoes originally cost $85.00, what was her total discount?

    A. 53.5%

    B. 46.5%

    C. 22.9%

    D. 39.2%

11. In July, gas prices increased by 15%. In August, they decreased by 10%. What is the total percent change since June?

    A. 5% increase

    B. 3.5% decrease

    C. 3.5% increase

    D. 1.5% increase

12. A bag contains twice as many red marbles as blue marbles, and the number of blue marbles is 88% of the number of green marbles. If g represents the number of green marbles, which of the following expressions represents the total number of marbles in the bag?

    A. $3.88 g$

    B. $3.64 g$

    C. $2.64 g$

    D. $2.32 g$

13. If 5 subtracted from 3 times x is greater than x subtracted from 15, which of the following is true of x?

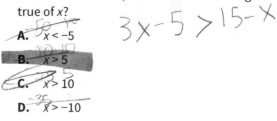

$3x - 5 > 15 - X$

    A. $x < -5$

    B. $x > 5$

    C. $x > 10$

    D. $x > -10$

14. Find the 10th term in the following sequence: 20, 8, −4, −16,...

    A. −100

    B. −88

    C. −72

    D. −136

15. If 38 is divided by m, the quotient is 12 and the remainder is 2. What is the value of m?

    A. 2

    B. 3

    C. 5

    D. 4

16. A radio station plays songs that last an average of 3.5 minutes and has commercial breaks that last 2 minutes. If the station is required to play 1 commercial break for every 4 songs, how many songs can the station play in an hour?

    A. 15

    B. 11

    C. 16

    D. 17

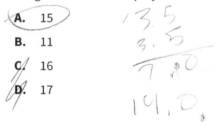

Word Knowledge

1. They investigated the <u>alleged</u> human rights violations.
   - A. proven
   - B. false
   - C. unproven
   - D. horrific

2. <u>Cede</u> most nearly means:
   - A. consign
   - B. surrender
   - C. keep
   - D. abandon

3. <u>Afflict</u> most closely means:
   - A. attack
   - B. perturb
   - C. assist
   - D. agonize

4. <u>Conspicuous</u> most nearly means:
   - A. bold
   - B. unremarkable
   - C. quiet
   - D. dull

5. <u>Insurgents</u> were responsible for a number of attacks, including suicide bombings.
   - A. anarchists
   - B. communists
   - C. rebels
   - D. patriots

6. <u>Austere</u> most nearly means:
   - A. welcoming
   - B. ornate
   - C. simple
   - D. fanciful

7. <u>Admonish</u> most closely means:
   - A. denounce
   - B. dislike
   - C. reprimand
   - D. praise

8. <u>Deference</u> most nearly means:
   - A. defiance
   - B. submissiveness
   - C. hostility
   - D. sociability

9. The site had been <u>neglected</u> for years.
   - A. ignored
   - B. maintained
   - C. crumbling
   - D. growing

10. <u>Insinuate</u> most closely means:
    - A. infiltrate
    - B. introduce
    - C. proclaim
    - D. abbreviate

11. <u>Explicate</u> most nearly means:
    - A. obscure
    - B. decipher
    - C. clarify
    - D. confuse

12. <u>Decorum</u> most nearly means:
    - A. propriety
    - B. decoration
    - C. drunkenness
    - D. indecency

13. He was <u>chagrined</u> when he tripped and fell in the hallway.

    a. injured

    b. embarrassed

    c. unharmed

    d. angry

14. <u>Audacious</u> most nearly means:

    a. frightening

    b. engaging

    c. daring

    d. boring

15. The <u>intrepid</u> volunteers worked in the refugee camps.

    a. uncaring

    b. caring

    c. compassionate

    d. fearless

16. <u>Surreptitious</u> most nearly means:

    a. hidden

    b. clandestine

    c. public

    d. illegal

Paragraph Comprehension

1. Every year the Academy Awards, or better known as The Oscars, brings together the best of the best in Hollywood. Each year since the original awards ceremony in 1929 great achievements in all areas of the film industry are recognized. Many married female actors, however, shy away from the honor of winning the *Academy Award of Merit* for either Best Actress or Best Supporting Actress. Ever since 1935, the "Oscar Curse" has proven more often than not to be alive and well. What is the "Oscar Curse" that these famous ladies of Hollywood fear?

   A. They fear that after winning they will meet an untimely end.

   B. That soon after winning this prestigious award, the lady's husband will leave them.

   C. The fear is that their next movie will be a box-office disaster.

   D. They fear that once they win one, they will never again win in the same category.

2. According to CNN.com, Google recently announced that it is developing smart contact lenses that will measure a diabetic's glucose level by testing the person's tears. If victorious, Google will eliminate a very laborious daily routine in every diabetic's life; drawing blood from their body (usually from the side of a finger) to test their glucose levels. In this paragraph, what does the word laborious mean?

   A. Consuming too much time

   B. Needing much unwelcome, often tedious, effort

   C. Needing to be done in a medical laboratory

   D. An excruciatingly painful procedure

3. Ikea stores have a unique section in their parking lots. They have a "family friendly" parking area. This area is located very close to the front entrance to the store. These spots have pink strollers painted on each parking spot. What is implied by the term "family friendly"?

   A. It is implying that only those customers who come to shop at the store with young children or pregnant women can park in this area.

   B. That if you have an Ikea Family Membership you are welcomed to park in this area.

   C. Any family, of any age, are welcome to park in this special area.

   D. That if there are only a few spots left in this area of the parking lot, it would be nice to leave it for a vehicle with a family but not it isn't necessary; anyone can park there.

4. Everyone dreams of winning the lottery; one million, 25 million, even 55 million dollars. It is very easy to get caught up in the dreams associated with winning the jackpot. The realists of the world, however, are quick to remind us that we have a better chance of being hit by a car than winning big with the lottery. What does the comparison of winning the lottery to being hit by a car imply?

   A. That if you don't have the good luck to win the lottery watch out because you only have bad luck and are likely to be hit by a car.

   B. It implies that it is not lucky to either win the lottery or be hit by a car.

   C. The comparison means that more people will get hit by a car than win big with the lottery.

   D. The implication is that if you are going to buy a lottery ticket, don't walk.

5. The United States Military Academy at West Point (USMA) is better known as The Point. Dating back to 1802, this coeducational federal service academy has trained some of the most revered and honored military leaders in American history. West Point has a Cadet Honor Code that is almost as old as the academy itself; "A Cadet will not lie, cheat, steal, or tolerate those who do." What is the foundation of the Honor Code of West Point?

A. The foundation of the Honor Code comes from a time when the United States where divided by the conflicts leading up to the American Civil War, but were training soldiers from both sides of the Mason-Dixie Line. This Code was required to prevent men from fighting amongst themselves.

B. This code came from the *Southern Gentleman's Guide to Behavior* and introduced to men from the northern states during the early years of the academy.

C. The Honor Code of West Point was adopted from the *British Military's Training Manual* that was created years before West Point even existed.

D. West Point's Code of Honor dates back to the beginning of the academy when a gentle man's word was considered his bond. To break one's word was the worst possible thing a gentleman could ever do. His word was his honor, and without honor a man was nothing.

6. Each branch of the United States Armed Forces has special mottos that the soldiers live and are expected to die by. These special expressions are points of extreme pride for each member of the military. What is the motto of the United States National Guard?

A. "This We'll Defend"

B. "Always Ready, Always There"

C. "That Others May Live"

D. "Not Self, but Country"

7. Davy Crockett is one of America's best-known folk heroes. Known for his political contributions to the State of Tennessee and the U.S. Congress, he also became famous during his own time for "larger than life" exploits that were retold through plays and in almanacs. Even following his death, Davy Crockett became growingly famous for exploits of legendary magnitude. In this paragraph, what is the meaning of the word *almanacs*?

A. An almanac is a book of information including a calendar, weather based predictions, anniversaries, and important events that is published yearly.

B. An almanac is another name for a book of locally developed plays that is published every couple years or so.

C. An almanac is a series of comics based on popular folklore that is published every five years.

D. An almanac is a name given to stories that are handed down from one generation to another orally, not by written word.

8. Rosa Parks was a civil rights activist who refused to give up her seat in the colored section on a city bus for a white person when the white section of the bus was full and was subsequently arrested. *My Story*, which is her autobiography, she is quoted as saying, "People always say that I didn't give up my seat because I was [physically] tired [or] old….No, the only tired I was, was tired of giving in." What is implied by this quote?

A. That she was old and tired of walking home after work each day and finally gave in and paid to take the bus home.

B. This quote implies that Rosa Parks was not tired physically, or too old to stand on a bus, she was just tired of having to give in to the demands of white people; she was tired of segregation based on race.

C. This quote means that people thought Rosa Parks was just too lazy to give up her seat on the bus.

D. Rosa Parks was just stubborn that day on the bus, and her actions had nothing to do with the civil rights movement.

9. When one wants to train a house-dog to ring a bell instead of barking to let its owner know it wants to go outside, there are only a few simple steps. First, when the dog is at the door, and barks take its paw and knock it against the bell that is hanging from the doorknob and only then open the door and let the dog outside. Repeat this every single time the dog barks to go outside. Eventually, depending on the stubbornness of the animal, the dog will cease barking at all and go to the bell and ring it each time it wants to go outside. What is the type of training called?

   A. This type of training is called Negative Behavior Elimination Training.

   B. This training is referred to as either Classical Conditioning or Pavlovian Conditioning.

   C. This training called Positive Reinforcement Training.

   D. This type of training is called Basic Cognitive Retraining.

10. When we think of rights we think in terms of Human Rights. This refers to ideas that apply to everyone, everywhere in the world. These expectations are egalitarian and are part of a declaration called the *Universal Declaration of Human Rights* that adopted by the U.N. General Assembly in 1948 after the end of WWII. In this paragraph, what does the word *egalitarian* mean?

   A. This word means that the rights contained in the *Universal Declaration of Human Rights* are to all be taken literally.

   B. Egalitarian means that ultimately these rights will also be applied to immediately to anyone and everyone who requests to be treated fairly.

   C. This word means that examples of basic human rights are included in the declaration adopted by the U.N.

   D. The word egalitarian means that Human Rights are the same for everyone, regardless of their race, nationality, or any other factors.

11. One island from the shores of San Francisco Bay is often referred to as "The Rock"; Alcatraz Island. The island has been home to one kind of prison or another since 1861 up until 1963. During its time as a federal prison, it is stated that no prisoner successfully escaped from Alcatraz although there were fourteen attempts in that time. Why were there never any successful escapes from the prison on Alcatraz Island?

   A. No one ever successfully escaped the prison because there were too many guards on duty. No man was ever left alone when outside of his cell.

   B. Alcatraz was inescapable because even if they penetrated the high-security around the prison, there was no way off the island since no boats were ever docked at the wharf.

   C. The entire premise of Alcatraz was that the men sent here were not to be rehabilitated back into society. Each and every aspect and component of the prison, the training of the guards, and the security around the rest of the island was created with the idea of keeping them on the island forever.

   D. The majority of men in the time the prison was active did not know how to swim, so those who attempted drowned in the water if they were not caught first.

GO ON

Math Knowledge

1. Which of these are parallel lines?

    A. $y = x + 5$, $y = -x + 5$

    B. $y = 2x + 3$, $y = 2x + 5$

    C. $y = 3x + 4$, $y = 2x + 4$

    D. $y = 4$, $x = 4$

2. Which of these are complementary angles?

    A. 63 degrees and 29 degrees

    B. 56 degrees and 38 degrees

    C. 33 degrees and 57 degrees

    D. 46 degrees and 49 degrees

3. The triangle whose one angle is greater than 90 degrees is called

    A. equilateral triangle

    B. isosceles triangle

    C. scalene triangle

    D. obtuse triangle

4. $a \times (b + c) =$

    A. $ab + bc$

    B. $cb + ac$

    C. $ab + ac$

    D. abc

5. Which of the following options is true for equilateral triangles?

    A. They have three congruent angles.

    B. They have three congruent sides.

    C. They have two congruent angles.

    D. They have two congruent sides.

6. If $\frac{20 - x}{4} = 3y$. What is x in terms of y?

    A. $20 - 12y$

    B. $20 + 12y$

    C. $12 - 20y$

    D. $12 + 20y$

    $20 \cdot x = 12y$

    $-x = 12y - 20$

    $x = -12y + 20$

7. For which of the following functions does $f(x) = |f(x)|$ for every value of x?

    A. $f(x) = 3 - x$

    B. $f(x) = 2x + x^2$

    C. $f(x) = x^3 + 1$

    D. $f(x) = x^2 + (2 - x)^2$

8. If $y = 7x$, and $x = 3z$, what will be the value of y if $z = 2$?

    A. 40

    B. 44

    C. 48

    D. 42

    $y = 7(3z)$

    $y = 7(6)$

    $y = 42$

    17

    $\frac{7}{99}$

    $\frac{70}{119}$

9. $4\frac{4}{6} + 2\frac{2}{3} - 1\frac{3}{4} \times 3\frac{2}{5} =$

    A. $\frac{22}{20}$

    B. $\frac{24}{20}$

    C. $\frac{21}{20}$

    D. $\frac{25}{20}$

    $\frac{28}{6} + \frac{14}{6} - \frac{7}{4} \cdot \frac{17}{5}$

    $\frac{119}{119}$
    $\frac{119}{357}$

    $\frac{920}{60} - \frac{119}{20}$

    $\frac{357}{63}$

10. Which one of the following options shows the correct answer of y with respect to its equation?

    A. If $2(y - 1) + 6 = 0$, then $y = 2$

    B. If $3(y - 3) = 3$, then $y = 4$

    C. If $2(y + 2) = 6$, then $y = -1$

    D. If $6y - 18 = 6$, then $y = 5$

    $4 - 2 + 6$

    $12 - 9$

    $\frac{63}{60} - \frac{21}{20}$

11. $A = x^2 + 3x - 4$, $B = 2x^2 - 2x + 3$. What will be the value of $B - A$?

    A. $x^2 - 5x + 7$

    B. $3x^2 - x - 1$

    C. $x^2 - 3x + 7$

    D. $x^2 - 5x - 7$

    $(2x^2 - 2x + 3) - (x^2 + 3x - 4)$

    $x^2 - 5x + 7$

12. The Pythagorean Theorem is applicable to which one of the following triangles?

    A. equilateral triangle

    B. acute triangle

    C. obtuse triangle

    D. right-angled triangle

13. $x = 3$ is the solution of which one of the following equations?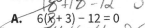

    A. $6(x + 3) - 12 = 0$

    B. $8(x - 2) - 4 = 0$

    C. $7(x - 6) + 21 = 0$

    D. $3(x + 4) - 9 = 0$

14. There are two parallel lines, x and y. One line s is passing through both these parallel lines such that $<smk = 60$ degrees. What will be the value of angle k?

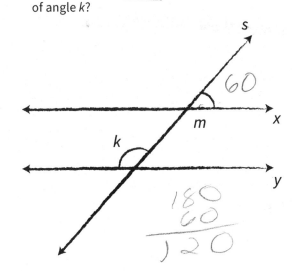

    A. 120 degrees

    B. 60 degrees

    C. 80 degrees

    D. 150 degrees

15. What will be the product of $3p^3 - 2p^2 + p$ and $-2p$?

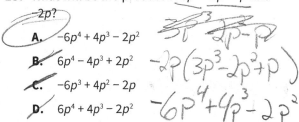

    A. $-6p^4 + 4p^3 - 2p^2$

    B. $6p^4 - 4p^3 + 2p^2$

    C. $-6p^3 + 4p^2 - 2p$

    D. $6p^4 + 4p^3 - 2p^2$

16. We have two numbers, x and y, such that $x + y = 15$, and $x - y = 3$. What will be the numbers?

    A. $x = 8, y = 5$

    B. $x = 10, y = 7$

    C. $x = 8, y = 7$

    D. $x = 9, y = 6$

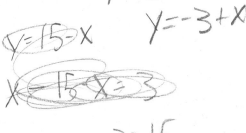

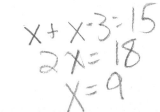

Electronics Information

1. Which one of the following circuits is a high-pass filter?

    A. RC differentiator

    B. RL integrator

    C. RC integrator

    D. all of the above

2. The capacitor which shorts an Alternating Signal (AC) to ground is called:

    A. motor capacitor

    B. bypass capacitor

    C. variable capacitor

    D. none of these

3. FET stands for which of the following?

    A. Ferroelectric Transistor

    B. Ferromagnetic-Efect Transistor

    C. Field-Effect Transistor

    D. Ferrocene-Electric Transistor

4. What is the relationship between voltage and current in case of a pure inductive circuit?

    A. Voltage is lagging Current by 90°

    B. Voltage is leading Current by 90°

    C. Current is leading Voltage by 90°

    D. Voltage is in phase with Current i.e. there is 0° angle between Current and Voltage

5. How many p-n junctions are there in a thyristor?

    A. one

    B. two

    C. three

    D. four

6. The terminals of a thyristor are:

    A. emitter, gate, collector

    B. anode, gate, cathode

    C. emitter, base, collector

    D. anode, base, cathode

7. Which one of the following options is not a region of operation of a BJT?

    A. forward-active mode

    B. saturation mode

    C. reverse-blocking mode

    D. both options B and C

8. Thyristor is also known as which of the following?

    A. silicon-controlled rectifier

    B. sodium-controlled rectifier

    C. sulfur-controlled rectifier

    D. silicon-controlled rectifier

9. In which of the following regions of operation, the transistor acts as a switching device?

    A. forward-active mode

    B. saturation mode

    C. cut-off mode

    D. both options B and C

10. Which of the following statements is correct regarding BJT?

    A. The emitter is heavily doped as compared to the collector.

    B. BJT has four terminals.

    C. While acting in forward-active mode, Collector-Base (C-B) junction is forward biased.

    D. Cut-off mode is mostly used for signal amplification.

11. Which one of the following statements is correct?

    A. MOSFET is a current-controlled device.

    B. BJT is a voltage-controlled device.

    C. Switching speed of a MOSFET is more than that of a BJT.

    D. both B and C

12. UPS stands for which of the following?

    A. Unidirectional Power Supply

    B. Uninterruptible Power Supply

    C. Unlimited Power Supply

    D. Unblocked Power Supply

13. The minority charge carriers in an n-type material are:

    A. electrons

    B. holes

    C. neutrons

    D. ions

14. How many electrons are present in the valence shell of a semi-conductor?

    A. two

    B. three

    C. four

    D. five

15. Which of the following statements is not correct regarding Zener Diode?

    A. Zener diode can be used as a voltage regulator.

    B. Zener diode can be used in the generation of reference voltage.

    C. Zener diode allows the current to flow in forward direction only.

    D. Zener diode can be used as an amplifier.

16. Two resistors of same value are connected in parallel to each other, the value of equivalent resistance will be:

    A. less than the value of one resistor

    B. greater than the value of one resistor

    C. the same as the value of one resistor

    D. equal to the sum of both resistors

GO ON

Auto and Shop Information

1. The pitman arm is a component of which automotive system?
   - **A.** suspension System
   - **B.** exhaust System
   - **C.** steering System
   - **D.** brake System

2. What controls the spark timing in some cars?
   - **A.** distributor
   - **B.** camshaft
   - **C.** crankshaft
   - **D.** timing Chain

3. If spark timing is advanced at high rpm, the spark takes place _____.
   - **A.** earlier in the combustion cycle
   - **B.** later in the combustion cycle
   - **C.** at the same time
   - **D.** at a different location in the combustion chamber

4. What opens and closes the exhaust and intake valves?
   - **A.** connecting Rods
   - **B.** crankshaft
   - **C.** camshaft
   - **D.** flywheel

5. What is the normal mixture of water and anti-freeze in an engine's coolant?
   - **A.** $\frac{30}{70}$
   - **B.** $\frac{50}{50}$
   - **C.** $\frac{70}{30}$
   - **D.** none of the above

6. Diesel engines have compression ratios between 14:1 and 23:1. Relative to gasoline engines, this is:
   - **A.** lower
   - **B.** higher
   - **C.** equal
   - **D.** none of the above

7. What splits power between the front and rear axles on a four-wheel drive vehicle?
   - **A.** transmission
   - **B.** torque converter
   - **C.** transfer Case
   - **D.** differential

8. How many degrees does the camshaft turn for a complete revolution of the crankshaft in a four-stroke engine?
   - **A.** 90
   - **B.** 180
   - **C.** 360
   - **D.** 720

9. What helps recharge the battery and run electrical components while the engine is running?
   - **A.** distributor
   - **B.** alternator
   - **C.** ignition System
   - **D.** starter

10. What system controls the ride quality of a vehicle?
    - **A.** suspension system
    - **B.** steering system
    - **C.** safety system
    - **D.** fuel delivery system

11. Which of the following oils would be the most viscous when starting in low temperature?

    A. SAE 10
    B. SAE 30
    C. 5W-20
    D. 10W-40

12. What is the length of a 10 – 32 × 1 bolt?

    A. 10 mm
    B. 32 mm
    C. 1 inch
    D. 10 inch

13. What should be used to strike a chisel or punch?

    A. rubber mallet
    B. ball peen hammer
    C. clawhammer
    D. all of the above

14. What type of hammer is used for assembling the frame of a house?

    A. finishing hammer
    B. sledgehammer
    C. deadblow hammer
    D. clawhammer

15. Which of the following is used for measuring a length?

    A. tape rule
    B. calipers
    C. steel rule
    D. all of the above

16. What is the diameter of a $\frac{1}{2}$ – 20 × 2 bolt?

    A. $\frac{1}{2}$ inch
    B. 20 mm
    C. 2 inch
    D. 2 mm

17. Which of the following can be used for shaping wood?

    A. chisel
    B. plane
    C. rasp
    D. all of the above

18. The value given by a pressure gauge is known as _____ pressure.

    A. absolute
    B. negative
    C. gage
    D. nominal

19. Which socket can be used with a $\frac{1}{2}$ inch drive impact wrench?

    A. $\frac{1}{4}$ inch flat black socket
    B. $\frac{1}{2}$ inch blue titanium socket
    C. $\frac{3}{4}$ inch chrome-plated socket
    D. $\frac{3}{4}$ inch flat black socket

20. Which sandpaper would leave the smoothest finish?

    A. 20 grit
    B. 100 grit
    C. 140 grit
    D. 200 grit

21. What joining process would be used to bond two small electrical wires together?

    A. brazing
    B. oxyacetylene welding
    C. soldering
    D. arc welding

22. Which of these would give the most accurate measurement of a small part?

    A. steel rule
    B. calipers
    C. micrometer
    D. tape measure

Mechanical Comprehension

1. Two moving bodies A and B possess the same amount of kinetic energy (see figure).

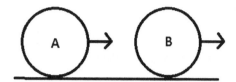

   If both the bodies are of unit mass then,

   A. velocity of body A > velocity of body B

   B. velocity of body A = velocity of body B

   C. velocity of body A < velocity of body B

   D. cannot be determined; insufficient data

2. A flywheel, initially at rest, attains an angular velocity of 600rad/s in 15sec. Assuming constant angular acceleration, the angular displacement and angular acceleration of the flywheel in this time is:

   A. 4500 rad, 40 rad/s^2

   B. 5400 rad, 40 rad/s^2

   C. 4000 rad, 45 rad/s^2

   D. 4000 rad, 54 rad/s^2

3. Calculate the mechanical advantage of the following wedge:

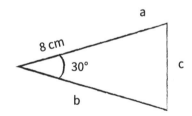

   A. one

   B. two

   C. three

   D. four

4. The wedge angle of a particular wedge is increased. The Mechanical Advantage (MA) of the wedge:

   A. increased

   B. decreased

   C. remained constant

   D. any of the possibilities is likely; MA not affected by the wedge angle

5. In the same question, the velocity of block Q, when it reaches a height of 7m above the ground is:

   A. 7.33 m/s

   B. 7.66 m/s

   C. 7.99 m/s

   D. cannot be determined; insufficient data

6. Observe the following figure:

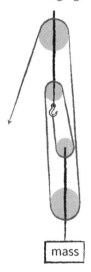

   The mechanical advantage (MA) for this frictionless pulley is:

   A. three

   B. four

   C. five

   D. cannot be determined; insufficient data

7. A body even after applying a certain amount of force, did not move. What can be said about the frictional force acting on the body?

   A. less than μmg

   B. more than μmg

   C. equal to mg

   D. equal to μmg

8. The gravitational force exerted by one object on another at macroscopic level:

   A. increases with the increase in distance

   B. decreases with the increase in distance

   C. remains constant

   D. none of the above

9. The dimensional formula of Gravitational constant is:

   A. ML^2T^{-2}

   B. $M^{-1}L^3T^{-1}$

   C. $M^{-2}L2T^{-2}$

   D. $M^{-1}L3T^{-2}$

10. A fighter jet traveling at a speed of 630 km/h drops a bomb 8 secs before crossing over a target to accurately hit the target. How far from the target was the jet when it dropped the bomb?

   A. 1.2 km

   B. 1.4 km

   C. 1.6 km

   D. 1.8 km

11. Which of the following is not an equation of uniformly accelerated motion:

   A. $a = \dfrac{v^2 - u^2}{2s}$

   B. $a = \dfrac{2(s - ut)}{t^2}$

   C. $a = \dfrac{2s - ut}{t^2}$

   D. $a = \dfrac{v - u}{t}$

12. Newton's 1st law of motion is based on the Galileo's law of inertia. Which of the following types of inertia satisfy this law:

   A. inertia of rest

   B. inertia of motion

   C. inertia of direction

   D. all of the above

13. Observe the figure below:

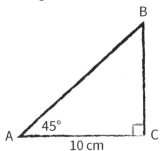

The mechanical advantage of the given ramp is:

   A. 1.414

   B. 0.141

   C. 14.14

   D. none of the above

14. Which of the following is not a part of the incline plane family of simple machines:

   A. wedge

   B. ramp

   C. lever

   D. screw

15. The mechanical advantage of a screw having 6 threads per inch and a radius of 0.1in is:

   A. 3.33

   B. 3.55

   C. 3.77

   D. 3.99

GO ON

16. Effort is being put on a lever with a speed of 20cm/s at a distance of 2m from the fulcrum. The speed at which the load moves, if it is located at a distance of 50cm from the fulcrum is:

A. 80 cm/s

B. 100 cm/s

C. 120 cm/s

D. 140 cm/s

Assembling Objects

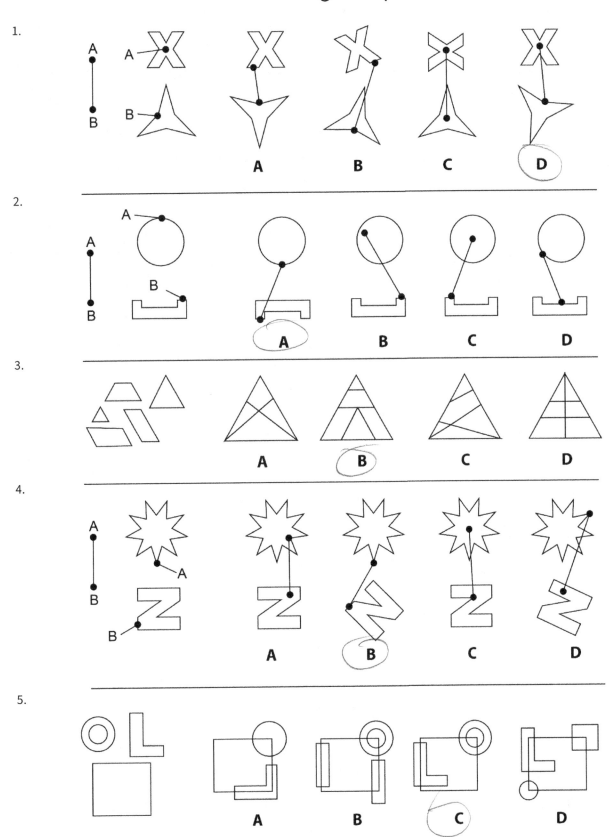

1.

    A   A   A   B   C   D

2.

    A   A   B   C   D

3.

                             A   B   C   D

4.

    A   A   B   C   D

5.

                             A   B   C   D

6.

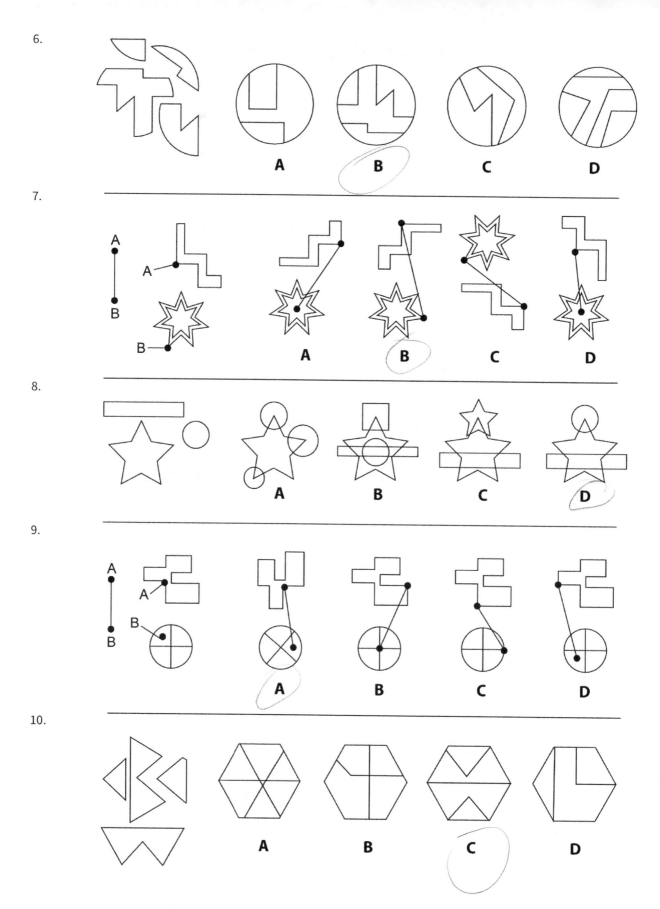

7.

8.

9.

10.

11.

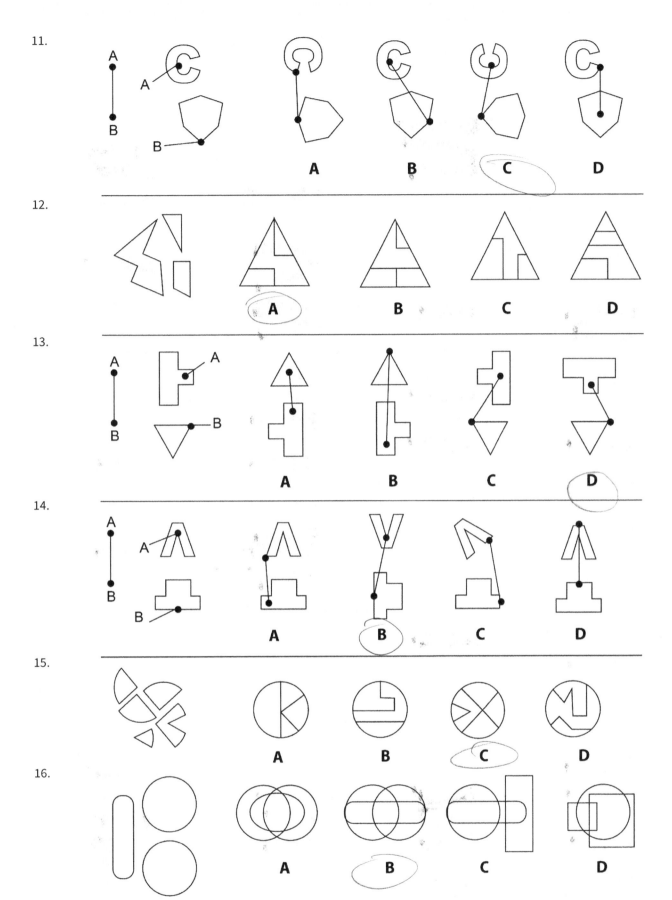

12.

13.

14.

15.

16.

Practice Test Three Answer Key

GENERAL SCIENCE

1. D.	5. D.	9. C.	13. D.
2. B.	6. A.	10. C.	14. B.
3. B.	7. B.	11. C.	15. C.
4. B.	8. A.	12. B.	16. B.

ARITHMETIC REASONING

1. C.	5. B.	9. C.	13. B.
2. B.	6. A.	10. B.	14. B.
3. A.	7. C.	11. C.	15. B.
4. D.	8. B.	12. B.	16. A.

WORD KNOWLEDGE

1. C.	5. C.	9. A.	13. B.
2. B.	6. C.	10. A.	14. C.
3. D.	7. B.	11. C.	15. D.
4. A.	8. B.	12. A.	16. B.

PARAGRAPH COMPREHENSION

1. B.	4. C.	7. A.	10. D.
2. B.	5. D.	8. B. ✓	11. C.
3. A.	6. B.	9. B. ✓	

MATHEMATICS KNOWLEDGE

1. B.	5. D.	9. C.	13. C.
2. C.	6. A.	10. B.	14. A.
3. D.	7. D.	11. A.	15. A.
4. C.	8. D.	12. D.	16. D.

ELECTRONICS INFORMATION

1. A.	5. C.	9. D.	13. B.
2. B.	6. B.	10. A.	14. C.
3. C.	7. C.	11. C.	15. C.
4. B.	8. A.	12. B.	16. A.

AUTO AND SHOP INFORMATION

1.	C.	7.	C.	13.	B.	19.	D.
2.	A.	8.	B.	14.	D.	20.	D.
3.	A.	9.	B.	15.	D.	21.	C.
4.	C.	10.	A.	16.	A.	22.	C.
5.	B.	11.	C.	17.	D.		
6.	B.	12.	C.	18.	C.		

MECHANICAL COMPREHENSION

1.	B.	5.	B.	9.	D.	13.	A.
2.	A.	6.	C.	10.	B.	14.	C.
3.	D.	7.	A.	11.	C.	15.	C.
4.	B.	8.	B.	12.	D.	16.	A.

ASSEMBLING OBJECTS

1.	D.	5.	C.	9.	A.	13.	D.
2.	A.	6.	B.	10.	C.	14.	B.
3.	B.	7.	B.	11.	C.	15.	C.
4.	B.	8.	D.	12.	A.	16.	B.

$$63\%$$

$$-\frac{54}{195}$$

sixteen

PRACTICE TEST FOUR

General Science

1. Which of the following is true regarding deoxyribonucleic acid (DNA). in the human body?

    A. DNA is used as an energy source.

    B. DNA is used as a template for creation of proteins.

    C. DNA is only found in the brain.

    D. DNA is made of sugar.

2. Testes are an organ found in:

    A. females

    B. plants

    C. males

    D. amoebas

3. How many kingdoms of life are there?

    A. three

    B. six

    C. seven

    D. nine

4. Plants absorb carbon dioxide (CO_2) to create sugar for energy. What is the primary byproduct of this process?

    A. oxygen

    B. nitrogen

    C. carbon monoxide

    D. carbon

5. What prevents ultraviolet radiation produced by the sun from damaging life on earth?

    A. the ozone layer

    B. greenhouse gasses

    C. the vacuum between earth and the sun

    D. the water layer

6. Which of the following is *not* present in an animal cell?

    A. nucleus

    B. mitochondria

    C. cytoplasm

    D. cell wall

7. Mitosis is the process of cell division to create new cells. What is the process of cell division required to create new sex cells or gametes?

    A. teloses

    B. meiosis

    C. kinesis

    D. phoresis

8. What are the two main parts of the human body's central nervous system?

    A. the heart and the spinal cord

    B. the brain and the spinal cord

    C. the peripheral nerves and the brain

    D. the spinal cord and the peripheral nerves

9. Which of the following is not an organ system in humans?

    A. the endocrine system

    B. the respiratory system

    C. the exophytic system

    D. the muscular system

10. Humans can turn glucose into ATP, the basic energy molecule in the body. What is a byproduct of this process?

    A. carbon dioxide

    B. oxygen

    C. nitrogen

    D. phosphorus

11. What is true of elements found in the same group (column) in the periodic table?

    A. They have the same atomic mass.

    B. They have the same level of reactivity.

    C. They have the same number of protons.

    D. The have the same number of valence electrons.

12. Compounds that are acidic will be able to lower the pH of a solution, by doing which of the following?

    A. accepting H+ ions

    B. releasing H+ ions

    C. binding with acidic species in solution

    D. reducing oxidative species in solution

13 Which of the following elements is the most electronegative?

    A. chlorine

    B. iron

    C. magnesium

    D. silicon

14. According to Newton's first law, $F = M \times A$, how fast will a 10-kilogram object accelerate when pushed with 50 Newtons of force?

    A. 2.5 m/s^2

    B. 5.0 m/s^2

    C. 8.0 m/s^2

    D. 15.0 m/s^2

    $50 = 10 \cdot a$

    5

15. How did mountains on the earth's surface primarily form?

    A. through the shifting of tectonic plates

    B. through the impact of meteors

    C. through gradual erosion

    D. through accumulation of soil by the wind

16. The age of the Earth is closest to:

    A. 100 million years

    B. 1.2 billion years

    C. 4.5 billion years

    D. 25.0 billion years

Arithmetic Reasoning

1. Order the following quantities on a number line, from most negative to most positive: $2^{-1}, -\frac{4}{3}, (-1)^3, \frac{2}{5}$

   A. $2^{-1}, -\frac{4}{3}, (-1)^3, \frac{2}{5}$

   B. $-\frac{4}{3}, (-1)^3, 2^{-1}, \frac{2}{5}$

   C. $-\frac{4}{3}, \frac{2}{5}, 2^{-1}, (-1)^3$

   D. $-\frac{4}{3}, (-1)^3, \frac{2}{5}, 2^{-1}$

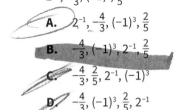

2. $\frac{8}{15}$ is $\frac{1}{6}$ of what number?

   A. $3\frac{1}{15}$

   B. $\frac{15}{48}$

   C. $\frac{4}{45}$

   D. $3\frac{1}{5}$

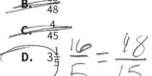

3. Jared finds a jacket in a store that is marked $\frac{1}{3}$ off. If his student discount gives him an additional $\frac{1}{5}$ off the original price, by what fraction is the jacket discounted in total?

   A. $\frac{1}{15}$

   B. $\frac{6}{15}$

   C. $\frac{7}{15}$

   D. $\frac{8}{15}$

   $\frac{3}{15} + \frac{5}{15}$

4. $0.003856 =$ 3.856×10^{-3}

   A. 3856×10^{-6}

   B. 385.6×10^{-5}

   C. 3.856×10^{-3}

   D. all of the above

5. Simplify: $[56 \div (2 \times 2^2)] - 9 \div 3$

   A. 4

   B. -0.667

   C. 34.33

   D. 109

   $[56 \div (8)] - 3$
   $7 - 3$

6. Becky is filling her rectangular swimming pool for the summer. The pool is 10 meters long, 6 meters wide, and 1.5 meters deep. How much water will she need to fill the pool?

   A. 90 meters

   B. 90 m²

   C. 90 m³

   D. 90 m⁴

7. Patrick is coming home from vacation to Costa Rica and wants to fill one of his suitcases with bags of Costa Rican coffee. The weight limit for his suitcase is 22 kilograms, and the suitcase itself weighs 3.2 kilograms. If each bag of coffee weighs 800 grams, how many bags can he bring in his suitcase without going over the limit?

   A. 27

   B. 23

   C. 4

   D. 2

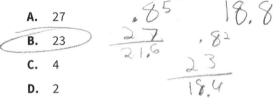

8. If m represents a car's average mileage in miles per gallon, p represents the price of gas in dollars per gallon, and d represents a distance in miles, which of the following algebraic equations represents the cost (c) of gas per mile?

   A. $c = \frac{dp}{m}$

   B. $c = \frac{p}{m}$

   C. $c = \frac{mp}{d}$

   D. $c = \frac{m}{p}$

9. Find the next term in the following sequence: 5, 12, 19, 26,...

   A. 35

   B. 37

   C. 33

   D. 34

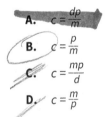

10. Suppose Mark can mow the entire lawn in 47 minutes, and Mark's dad can mow the entire lawn in 53 minutes. If Mark and his dad work together (each with their own lawnmowers), how long will it take them to mow the entire lawn?

$\frac{1}{2} = 23.5$
$= 26.5$

A. 15.6 minutes

B. 24.9 minutes

C. 26.5 minutes

D. 50 minutes

11. The kinetic energy, K, of an object is given in terms of its mass, m, and velocity, v, as shown in the equation $K = \frac{1}{2}mv^2$. Object A and object B have the same mass. If object A is moving at half the velocity of object B, the kinetic energy of object A is what fraction of the kinetic energy of object B?

A. $\frac{1}{2}$

B. $\frac{1}{4}$

C. $\frac{1}{16}$

D. $\frac{1}{64}$

12. Stephanie eats 0.625 of her pizza. If her pizza was cut into 8 slices, how many slices has she eaten?

$\begin{array}{r} 8 \\ \times .625 \\ \hline 5.000 \end{array}$

A. 3

B. 4

C. 5

D. 6

13. Gym A offers a monthly membership for 80% of the cost at Gym B; the cost at Gym B is 115% the cost of Gym C. What percentage of the cost at Gym C does Gym A charge?

A. 35%

B. 97%

C. 70%

D. 92%

$\begin{array}{r} \times 115 \\ \hline 80 \\ \hline 35 \end{array}$

14. Simplify: $54.48 \div 0.6$

A. 0.908

B. 9.08

C. 90.8

D. 908

$\begin{array}{r} 54.48 \\ .6 \\ \hline \end{array}$

$90.8 \quad 59.48$

15. John and Jake are working at a car wash. It takes John 1 hour to wash 3 cars; Jake can wash 3 cars in 45 minutes. If they work together, how many cars can they wash in 1 hour?

A. 6 cars

B. 7 cars

C. 9 cars

D. 12 cars

16. Melissa is ordering fencing to enclose a square area of 5625 square feet. How many feet of fencing does she need?

A. 75 feet

B. 150 feet

C. 300 feet

D. 5,625 feet

$\begin{array}{r} 375 \\ 75 \\ \hline 375 \\ 5250 \\ \hline 5525 \end{array}$

$\begin{array}{r} 150 \\ 150 \\ \hline 000 \\ 750 \\ \hline 22500 \end{array}$

Word Knowledge

1. The course was <u>required</u>.
   - **A.** mandatory
   - **B.** voluntary
   - **C.** managed
   - **D.** volatile

2. <u>Stalwart</u> most closely means:
   - **A.** weak
   - **B.** ill
   - **C.** brave
   - **D.** stubborn

3. His responses and personality were <u>brusque</u>.
   - **A.** cheerful
   - **B.** slow
   - **C.** curt
   - **D.** angry

4. <u>Subordinate</u> most nearly means:
   - **A.** lower
   - **B.** higher
   - **C.** larger
   - **D.** smaller

5. The test caused him to feel <u>apprehensive</u>.
   - **A.** nervous
   - **B.** terrified
   - **C.** excited
   - **D.** elated

6. <u>Proscribe</u> most closely means:
   - **A.** prohibit
   - **B.** prescribe
   - **C.** require
   - **D.** write

7. <u>Infallible</u> most nearly means:
   - **A.** flawed
   - **B.** broken
   - **C.** inaccurate
   - **D.** flawless

8. He was <u>meticulous</u> about his appearance in uniform.
   - **A.** thorough
   - **B.** conscientious
   - **C.** lackadaisical
   - **D.** pretentious

9. She was <u>deferential</u> to her commanding officers.
   - **A.** disrespectful
   - **B.** respectful
   - **C.** rude
   - **D.** pleasant

10. <u>Comply</u> most nearly means:
    - **A.** hurry
    - **B.** follow
    - **C.** ignore
    - **D.** obey

11. <u>Deploy</u> most nearly means:
    - **A.** post
    - **B.** fight
    - **C.** ship
    - **D.** return

12. <u>Inept</u> most closely means:
    - **A.** clumsy
    - **B.** skilled
    - **C.** conscientious
    - **D.** incompetent

13. He was a <u>conscientious</u> worker.

   A. careless

   B. sloppy

   C. careful

   D. slow

14. <u>Malevolent</u> most closely means:

   A. bad

   B. good

   C. slow

   D. fast

15. <u>Treacherous</u> most nearly means:

   A. illegal

   B. dangerous

   C. unhealthy

   D. immoral

16. He was <u>irate</u> when his instructions were not followed to the letter.

   A. irritated

   B. frustrated

   C. enraged

   D. calm

Paragraph Comprehension

1. Between April 1860 and October 1861 The Pony Express delivered mail, news, and other forms of communication from Missouri across the Great Plains, through the Rocky Mountains, through the desert lands of Nevada to California, using only man and horse power. The Pony Express closed in October of 1861; just two days after the transcontinental telegraph reached Salt Lake City, therefore, connecting Omaha, and Nebraska to California. Other telegraph lines connect many other cities along the Pony Express Route. Why did the Pony Express close?

   A. The Civil War stopped them from running their business.

   B. Another company was faster and took over the business.

   C. The Pony Express riders were unable to pass through the Rocky Mountains in the winter months.

   D. With the transcontinental telegraph connecting so many cities along the route, the Pony Express became redundant.

2. Between 1914 and 1935, George Herman "Babe" Ruth Jr. was known as "the Bambino" to baseball fans. Over his 22 seasons, he only played for three teams (Boston Red Sox, New York Yankees, and Boston Braves) and was known most for his hitting skills and RBI's statistics. Due mostly to Babe Ruth's hitting ability baseball changed during the 1920's from a fast-playing game with lower scores to one of higher scores and a slower pace. How did "the Bambino's" hitting skills and RBI's statistics affect the way baseball was played?

   A. He hit so many batters in that the game went faster.

   B. The innings lasted longer with so many batters scoring runs.

   C. They had to stop the game because every time Babe Ruth hit a home run fans mobbed him.

   D. The Regulations changed which caused the game to last longer.

3. Kraft Macaroni and Cheese goes by many names. In Canada, it is called Kraft Dinner and in the United Kingdom it is known as Cheesy Pasta. No matter what name it is called by, this pasta dish has been a staple of the typical North American diet since its beginning in 1937. James Lewis Kraft, a Canadian living in Chicago struck gold by introducing this product during WWII, when more and more women were working outside of the home, milk and other dairy foods were rationed and hearty "meatless" meals were relied upon. Why has this product continued to be a staple in our diet over 75 years after it was introduced to Americans?

   A. Most Americans love pasta and cheese.

   B. It is still the cheapest pasta on the market.

   C. The same factors that made its introduction so popular still exist today.

   D. It is still popular today because of brilliant marketing strategies.

4. Nelson Mandela, Steve Biko, Desmond Tutu, Denis Goldberg, and Harry Schwarz are all activists that fought against the system of apartheid in countries such as South Africa during the 1980's and 1990's. What is the meaning of the word "Apartheid"?

   A. Apartheid is a system of segregation based on a person's race that was law in South Africa between 1948 and 1994.

   B. Apartheid is a system by which people in South Africa are dictated where they are allowed to live and who they can marry based on how much wealth their family has accrued.

   C. Apartheid is the name given to the trade embargo placed on the country of South Africa by the United States of America between 1950's and 1990's.

   D. Apartheid is a system of classification South Africa, based on the race of your grandparents; therefore, every second generation is able to apply for a change in classification based on inter-racial changes in the family tree.

5. The *Declaration of Independence* was unanimously voted on July 2 and adopted by the Continental Congress on July 4, 1776; declaring considered themselves no longer one of Britain's colonies. The official announcement was also made on July 4th to the people of the original thirteen American "states" that made up the new country known as the United States of America. Why is Independence Day celebrated on July 4th and not July 2nd when the unanimous vote occurred?

A. Since July 4th was the first weekday that the American people heard the announcement of the *Declaration of Independence*, that is the day that people counted as the first day as an independent country.

B. Officially, until the *Declaration of Independence* was adopted by the Continental Congress and announced to the people of the thirteen colonies, it was not considered legally binding.

C. Legally the *Declaration of Independence* should be observed on July 2; the people of the U.S.A. are wrong.

D. The politicians involved in the original signing of the *Declaration of Independence* didn't want their new country's birthday to be the day after Canada celebrates the date they became a country independent of the British Empire.

6. The RMS Titanic, although built in Belfast, Ireland, was a British passenger ship that collided with an iceberg in the North Atlantic on April 15, 1912, and sank to the bottom of the ocean. More than 1,500 souls died in what is considered one of the deadliest maritime disasters outside of war. The ocean liner only had enough lifeboats for just over half of the number of people on board. In this paragraph, what does RMS stand for?

A. Royal Majesty's Service

B. Royal Mail Ship

C. Royal Majesty's Ship

D. Royal Mail Steamer

7. The Statue of Liberty was a gift from the people of France and was dedicated on October 28, 1886. She is an iconic symbol of patriotism to the people of the United States of America and even has the date of the American Declaration of Independence, July 4, 1776, chiseled on the tablet she is holding. Displayed on Liberty Island, this statue is a beacon of hope and freedom to immigrants from around the world. Why is the Statue of Liberty a symbol of hope and freedom to immigrants from around the world?

A. The Statue of Liberty is the very first site that any immigrant traveling by boat sees when entering the New York Harbor. She is the epitome of everything the United States of America stands for, "Life, Liberty, and the pursuit of Happiness."

B. The Statue of Liberty has been the only symbol of the United States that has been used in films and other forms of media that people in other countries have seen and, therefore, associates it with the country.

C. The Statue of Liberty was a sign of peace sent from France after they lost the Great War to the United States. It was a sign of freedom because the U.S.A did not conquer their country when they won the war and a sign of hope that peace will reign between the two countries.

D. Immigrants know that once they have passed the Statue of Liberty, under no circumstances will they be forced to return to their home country and lose the freedom of living in the United States of America.

8. The "Anti-Fascist Protection Rampart" was the official name given to the Berlin Wall. This wall was a large concrete wall with guard towers that completely blocked West Berlin from both East Germany and East Berlin. It was claimed by the German Democratic Republic (GDR), the political party in power after the end of WWII in Eastern Germany, that this wall was built to protect the people from the fascist interests attempting to prevent East Berliners from building a socialist state which was their idea of utopia. Between the years of 1961 and 1989 it is estimated that 5000 people attempted to escape over the wall resulting in over 100 deaths. What is implied by the fact that so many people "attempted to escape" over the Berlin Wall?

A. It is implied that many Western Germans wanted to live in Eastern Germany and live in their utopia county.

B. It is implied that Eastern Germany was very "fore-thinking" in their policies for national security although, a little over-the-top compared to today's standards.

C. Although the political party believed in "protecting" the people of East Germany, many East German's were not in agreement with the "utopia" world that was developed but were not free to just leave.

D. East Germans had to pay a very high tax in order to be able to move out of the area, so many people tried to leave without having to pay so much money to the government.

9. A tropical cyclone is a storm system that rotates very quickly and is pigeonholed by strong winds, heavy rain, and a low-pressure center. Some names that a tropical cyclone is referred to as include: typhoon, hurricane, and tropical depression. In this paragraph, what does the word "pigeonholed" means?

A. A small compartment that is part of a set on a desk or wall unit into which things are placed.

B. A small hole in a tree where birds build a nest.

C. A hole in a piece of wood that is roughly the size of a pigeon.

D. A broad category or label given to someone or something.

10. Mickey Mouse is the official mascot of The Walt Disney Company and was first created in 1928 by Walt Disney and Ub Iwerks as a roguish antihero. Since his first appearance in test screening of Plane Crazy, Mickey Mouse has appeared in over 130 films, comic strips, comic books, video games, television series, and is now a lovable, friendly character that can be seen at all of the Disney Theme Parks. Why has Mickey's character changed from antihero to fun-loving and adventurous over the last 80 years or so?

A. Mickey Mouse scared children when he was first introduced as an impish antihero and Walt Disney was forced to change his characteristics or create another character to be the official mascot.

B. Although Mickey Mouse was first created as an adult character, Walt Disney realized that more and more children were watching the movies, therefore, he altered his image to be more "child-like" so that his younger audience could identify with him.

C. As Mickey Mouse's popularity grew, and more characters were developed to be his friends, such as Minnie Mouse, Pluto, and Donald Duck, it was only natural that his personality changed into a friendlier, fun-loving character.

D. Walt Disney received thousands of letters from parents complaining that their children were misbehaving and acting out the behaviors of the mischievous mouse featured in the Disney films and comics. Walt Disney was pressured by the public to change Mickey Mouse's characteristics, or America's parents would have boycotted the entire Disney brand.

GO ON

11. Historically, archeologists have identified many dinosaurs that are omnivores. This means that eat both plants and meat. Other dinosaurs are only plant-eaters or meat-eaters; they are referred to as herbivores and carnivores, respectively. What is the main idea in this paragraph?

  A. The main idea is that there were herbivore, carnivore, and omnivore dinosaurs on this planet thousands of years ago.

  B. The main idea of this paragraph is that archeologists have identified different types of dinosaurs.

  C. Herbivores were dinosaurs that ate only a diet of plants; this is the main idea of the paragraph.

  D. The main idea of this paragraph is that carnivores were dinosaurs that only ate other animals.

Mathematics Knowledge

1. $\frac{4}{5} \div \frac{6}{7} \times \frac{1}{2} + \frac{3}{2}$

A. $\frac{56}{30}$

B. $\frac{57}{30}$

C. $\frac{58}{30}$

D. $\frac{59}{30}$ (circled)

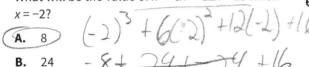

(handwritten work): $\frac{4}{5} \cdot \frac{7}{6} = \frac{28}{30} \cdot \frac{1}{2} = \frac{28}{60} = \frac{14}{30}$ $\frac{14}{30} + \frac{45}{30} = \frac{59}{30}$

2. What will be the value of $x^3 + 6x^2 + 12x + 16$ when $x = -2$?

A. 8 (circled)

B. 24

C. 48

D. 72

(handwritten work): $(-2)^3 + 6(-2)^2 + 12(-2) + 16$ $= -8 + 24 - 24 + 16 = 8$

3. ABCD is a rectangle and inside it ABE is an Equilateral Triangle. What will be the angle ∡CEA represented by x in the figure?

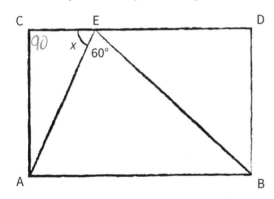

A. 45 degrees

B. 50 degrees

C. 60 degrees (circled)

D. 55 degrees

4. Which of the following numbers will satisfy this equation $\frac{n(n+1)(n+2)}{n(n+4)} = 2$

A. 1

B. 2 (circled)

C. 3

D. 4

(handwritten work): $\frac{n(n^2 + 3n + 2)}{n + 4}$

5. If $6x^2 + 7y = 45$ is an equation. What will be the value of y if $x = 2$?

A. 4

B. 1

C. 2

D. 3 (circled)

(handwritten work): $6(2)^2 + 7y = 45$ $24 + 7y = 45$ $7y = 21$ $y = 3$

6. The figure shows a right-angled triangle with side AC = 5m. Find the length of side AB.

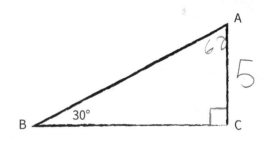

A. 7

B. 8

C. 9

D. 10 (circled)

7. If $z = \frac{6y^2}{3x}$, what will be the value of z for $y = 3x$ and $x = 2$?

A. 36 (circled)

B. 30

C. 32

D. 40

(handwritten work): $z = \frac{6(3x)^2}{3x} = \frac{6(9x^2)}{3x}$ $\frac{54x^2}{3x} = z = \frac{6(6)^2}{3(2)} = \frac{6(36)}{6}$

8. If we divide the first number by the second number, we get 3, and the sum of both numbers is 16. What will be the numbers?

A. 15, 5

B. 12, 4 (circled)

C. 9, 3

D. 18, 6

$y = mx + b$

9. What will be the point of intersection of two lines $3x + 2y - 8 = 0$ and $4x + 7y - 15 = 0$?

$2y = -3x + 8$
$y = \frac{3}{2}x + 4$

A. $(1, 2)$
B. $(2, 1)$
C. $(1, 3)$
D. these are non-intersecting lines.

$7y = -4x + 15$
$y = \frac{-4x}{7} + \frac{15}{7}$

10. How many prime numbers are less than 29?

A. 7
B. 8
C. 9
D. 10

$2, 3, 5, 7, 11, 13,$
$17, 23, 26$

11. 5,500,000 is not equal to which of the following?

A. 5.50×10^6
B. 550×10^4
C. 55×10^6
D. 0.55×10^7

5.5×10^6

12. What is the correct order if the following fractions are placed in order from largest to smallest? $\frac{1}{3}, \frac{1}{2}, \frac{3}{4}, \frac{5}{6}$

$\frac{4}{12}, \frac{6}{12}, \frac{9}{12}, \frac{10}{12}$

A. $\frac{1}{2}, \frac{5}{6}, \frac{3}{4}, \frac{1}{3}$
B. $\frac{5}{6}, \frac{1}{2}, \frac{1}{3}, \frac{3}{4}$
C. $\frac{3}{4}, \frac{1}{3}, \frac{1}{2}, \frac{5}{6}$
D. $\frac{5}{6}, \frac{3}{4}, \frac{1}{2}, \frac{1}{3}$

13. $-(-5)^3 \div (-5)^2 =$

$125 \div 25$

A. -125
B. 25
C. 125
D. -25

14. $\left(\frac{2a^{10}b^5}{3}\right) \div \left(\frac{15a^2b}{2a^{-3}}\right)^{-1} =$

$\frac{15^{-1}a^{1}b^{0}}{2^{-1}a^{4}}$

A. $\frac{4a^5b^6}{45}$
B. $5a^{15}b^5$
C. $5a^{12}b^7$
D. $5a^{15}b^7$

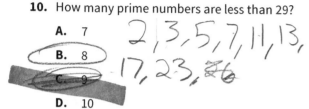

$2a^{10}b^6 \cdot \frac{2a^4}{15ab}$

$\frac{4a^{13}b^6}{45a^2b} \quad \frac{4a^{14}b^6}{45ab}$

15. If the sum of the digits of a number is 11, the product of its digits is 16, and the difference of the tenth unit digit from hundredth unit digit is 6. What is the number?

A. 821
B. 812
C. 218
D. 128

16. If $x > y$, $y < z$, $x > z$, such that product of x, y, and z is 72. What are the numbers?

A. $x = 6, y = 4, z = 3$
B. $x = 4, y = 6, z = 3$
C. $x = 3, y = 4, z = 6$
D. $x = 6, y = 3, z = 4$

Electronics Information

1. The minimum amount of energy required to disclose an electron from a covalent bond is called which of the following?

   A. transition energy

   B. bandgap energy

   C. thermal energy

   D. potential energy

2. An intrinsic semi-conductor is a semi-conductor in which:

   A. No significant dopant species are present, and it is a pure semi-conductor.

   B. A doping agent has been added to change its properties.

   C. The concentrations of electrons and holes change due to doping.

   D. The resistivity of the semi-conductor can change by adding impurities.

3. Which of the following statements is correct regarding the flow of current?

   A. The current flows from low potential to high potential.

   B. The current flows in the opposite direction of electric field.

   C. The current flows from high potential to low potential.

   D. The current flow does not depend on the potential difference.

4. The capacitor blocks Direct Current (DC).Hence, it is also known as:

   A. blocking capacitor

   B. switched capacitor

   C. coupling capacitor

   D. both A and C

5. According to Ohm's Law, what will happen to the resistance if the current increases?

   A. Resistance will increase.

   B. Resistance will decrease.

   C. Resistance will remain same.

   D. Resistance will become zero.

6. The resistance of a wire is R. What will be the resistance if the length of the wire becomes double, and the cross-sectional areas become half?

   A. $\left(\frac{1}{4}\right)R$

   B. R

   C. 2R

   D. 4R

7. What is the equivalent capacitance of four capacitors connected in parallel if the value of each capacitor is 4F?

   A. 1F

   B. 10F

   C. 16F

   D. 20F

8. What are two terminals of a diode?

   A. collector, cathode

   B. anode, cathode

   C. anode, emitter

   D. emitter, collector

9. The diode which is designed to go through avalanche breakdown at a specified reverse bias voltage is called:

   A. light-emitting diode

   B. avalanche diode

   C. varactor diode

   D. switching diode

10. Which one of the following is not a region of operation of a diode?

A. forward-bias region

B. breakdown region

C. saturation region

D. reverse-bias region

11. BJT stands for which of the following?

A. Biased Junction Transistor

B. Bisection Junction Transistor

C. Bidirectional Junction Transistor

D. Bipolar Junction Transistor

12. What is the function of an inductor?

A. storing current

B. controlling current

C. acting as a non-passive device

D. creating potential difference

13. What is the function of an inverter?

A. It converts AC to DC.

B. It steps up AC.

C. It converts DC to AC.

D. It steps up DC.

14. What is the function of a transformer?

A. It transforms voltage.

B. It transforms current.

C. It transforms power.

D. It transforms both voltage and current.

15. Which one of the following is a step-down DC to DC converter?

A. boost converter

B. buck converter

C. transformer

D. inverter

16. Power factor is represented by which of the following?

A. $\cos \theta$

B. $\sin \theta$

C. $\tan \theta$

D. $\cot \theta$

Auto and Shop Information

1. Which of the following ignition systems has contact points?

    A. distributorless ignition

    B. direct ignition

    C. capacitor discharge ignition

    D. none of the above

2. The catalytic converter is a component of which automotive system?

    A. emission control system

    B. drivetrain

    C. safety system

    D. cooling system

3. What can be the side effects of worn piston rings?

    A. increased oil consumption

    B. white or gray smoke

    C. low power

    D. all of the above

4. How many times does the crankshaft rotate during a complete cycle of a four-stroke engine?

    A. one

    B. two

    C. three

    D. four

5. What is the most common type of automotive engine?

    A. rotary

    B. four cycle

    C. two cycle

    D. external combustion

6. If an engine is being supplied with too much fuel, the air-fuel mixture is said to be:

    A. rich

    B. lean

    C. stoichiometric

    D. high octane

7. What gauge shows the speed of an engine in rpm?

    A. speedometer

    B. tripometer

    C. tachometer

    D. fuel gauge

8. Most modern automotive brake systems are activated:

    A. mechanically

    B. electronically

    C. hydraulically

    D. pneumatically

9. How is fuel mixed into the airstream entering a carbureted engine?

    A. low pressure in the venturi of the carburetor

    B. injectors controlled by the Power Control Module (PCM)

    C. high pressure in the fuel line created by the fuel pump

    D. both A and C

10. In a gasoline engine, an air-fuel mixture of 11:1 would be considered:

    A. rich

    B. lean

    C. stoichiometric

    D. none of the above

11. A car's radiator transfers heat from the
_____ to the air.

    A. engine oil

    B. distributor

    C. coolant

    D. brake fluid

12. Which tool can be used to measure an angle?

    A. compass

    B. protractor

    C. dial indicator

    D. calipers

13. Which of the following cuts with a pulling
motion?

    A. backsaw

    B. file

    C. pull saw

    D. ripsaw

14. Which of these is not a common type of file?

    A. flat

    B. half-round

    C. hexagon

    D. triangle

15. What tool would be used to tighten a bolt to
20 ft-lb?

    A. torque wrench

    B. breaker bar

    C. pipe wrench

    D. box wrench

16. Which tool is used to remove a small strip from
wood?

    A. file

    B. ripsaw

    C. plane

    D. chisel

17. Which of these saws creates the widest kerf?

    A. crosscut saw

    B. jigsaw

    C. ripsaw

    D. hacksaw

18. What is the pitch of an M10 × 1.25 × 40 bolt?

    A. 10 mm

    B. 1.25 tpi

    C. 40 tpi

    D. 1.25 mm

19. Which of the following should be used to cut
wood along the grain?

    A. crosscut saw

    B. ripsaw

    C. backsaw

    D. knife

20. Which fractional screw has the finest thread?

    A. $\frac{1}{4} - 20 \times \frac{3}{4}$

    B. $10 - 32 \times 1$

    C. $\frac{7}{16} - 20 \times 2$

    D. $\frac{1}{2} - 24 \times 3$

21. Which of these tools is considered a fastening
tool?

    A. stapler

    B. screwdriver

    C. plier

    D. all of the above

22. Which of the following is a thread type?

    A. unified national coarse

    B. unified national wide thread

    C. unified national fine

    D. both A and C

Mechanical Comprehension

1. Calculate the amount of work done in moving a mass of 10kg at rest with a force of 5N in 8 seconds with no repulsive forces in action?

   A. 80J

   B. 100J

   C. 120J

   D. 60J

2. Consider 3 equal masses at arbitrary points A, B and C in space and let D be a point on the surface of the earth (as shown in the figure).

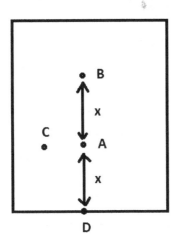

Then,

> **i.** The mass at point B has the maximum potential energy
>
> **ii.** Masses at points A and C have equal P.E. but less than that of the mass at point D
>
> **iii.** The mass at point D, if lifted to a height 2x, will possess P.E. equal to P.E.(B).

   A. Statements (i), (ii) and (iii) are true.

   B. Only statements (i) and (iii) are true.

   C. None of them is true.

   D. Only statements (i) is true.

3. The disc in the figure with mass m is set to roll with angular velocity omega.

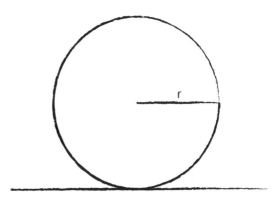

Total energy of the ball is:

   A. $\frac{1}{2}mr^2w^2$

   B. $\frac{3}{4}mrw^2$

   C. $\frac{1}{2}mrw^2$

   D. $\frac{3}{4}mr^2w^2$

4. Consider the following table:

S.No.	Length (cm)	Breadth (cm)
wedge 1	5	2
wedge 2	7	3
wedge 3	4	6
wedge 4	9	4

Which of the above wedges will provide maximum Mechanical Advantage?

   A. wedge 1

   B. wedge 2

   C. wedge 3

   D. wedge 4

5. Acceleration of a moving body can be determined by:

   A. slope of velocity-time graph

   B. area under velocity time graph

   C. slope of distance-time graph

   D. area under distance-time graph

6. A block P of mass 100kg is tied to one end of a rope on a frictionless pulley to lift another block Q with mass 40kg. What is the amount of acceleration produced in the rope during the lifting action?

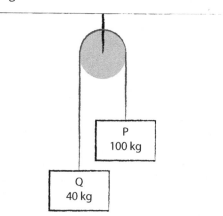

    A.   2.4 m/s²

    B.   3.3 m/s²

    C.   4.2 m/s²

    D.   5.1 m/s²

7. In the above problem, the value of acceleration for block P and Q are:

    A.   negative and positive respectively

    B.   positive and negative respectively

    C.   both negative

    D.   both positive

8. A train of mass 100 tons is traveling with a uniform velocity of 108kmph. The driver is informed about a broken bridge 250m away from his present location. He immediately applies the brakes. If the coefficient of friction between the train and the rails is 0.2, what is the uniform deceleration achieved by the train after the brakes were applied?

    A.   1.88 m/s²

    B.   1.92 m/s²

    C.   1.96 m/s²

    D.   2.00 m/s²

9. In the previous problem, was the driver able to prevent the accident?

    A.   yes

    B.   no

    C.   cannot be determined

    D.   none of the above

10. Ratio of the distance traveled by a free falling body in the 1st 3 seconds is:

    A.   1:4:8

    B.   1:2:3

    C.   1:7:11

    D.   1:4:9

11. Which type of simple machine is used in each case?

i.	woodcutter using an ax
ii.	lady withdrawing water from well
ii.	children playing on see-saw
iv.	a mechanic working with nuts and bolts

    A.   screw, wedge, pulley, lever

    B.   wedge, pulley, lever, screw

    C.   pulley, lever, screw, wedge

    D.   lever, screw, wedge, pulley

12. Which of the following statements is true:

    A.   IMA is always > AMA

    B.   IMA is always < AMA

    C.   IMA is always = AMA

    D.   none of the above

13. In wrestling matches, soft ground is provided instead of hard ground because:

    A.   During a fall, frictional force can cause burns to wrestler.

    B.   Hitting ground is an impulsive force.

    C.   Soft ground provides better recoil.

    D.   none of the above

14. Which of the following statements is true:

> **i.** FE is in the middle of the 3rd class lever
>
> **ii.** Fulcrum is in the middle of the 1st class lever
>
> **iii.** FR is in the middle of the 2nd class lever

- **A.** Only statement (i) and (ii) are correct.
- **B.** All statements are correct.
- **C.** Only statement (ii) is correct.
- **D.** None of the statements is correct.

15. A 100 kg stone and a bird feather are allowed to free fall in the vacuum from a height of 50m. The time taken by 100 kg stone to reach ground is:

- **A.** less than the time taken by the feather
- **B.** greater than the time taken by the feather
- **C.** equal to the time taken by the feather
- **D.** infinite as there is no force exerted because the medium is vacuum

16. Consider the following statements:

> **i.** Only order 1 and order 2 levers multiply force
>
> **ii.** Mechanical Advantage of 3rd order lever is always < 1

- **A.** Statement (i) is true, but statement (ii) is false.
- **B.** Statement (i) is false, but statement (ii) is true.
- **C.** Both statements are true, but statement (ii) is not the correct explanation of statement (i).
- **D.** Both statements are true and statement (ii) is the correct explanation of statement (i).

GO ON

Assembling Objects

1.

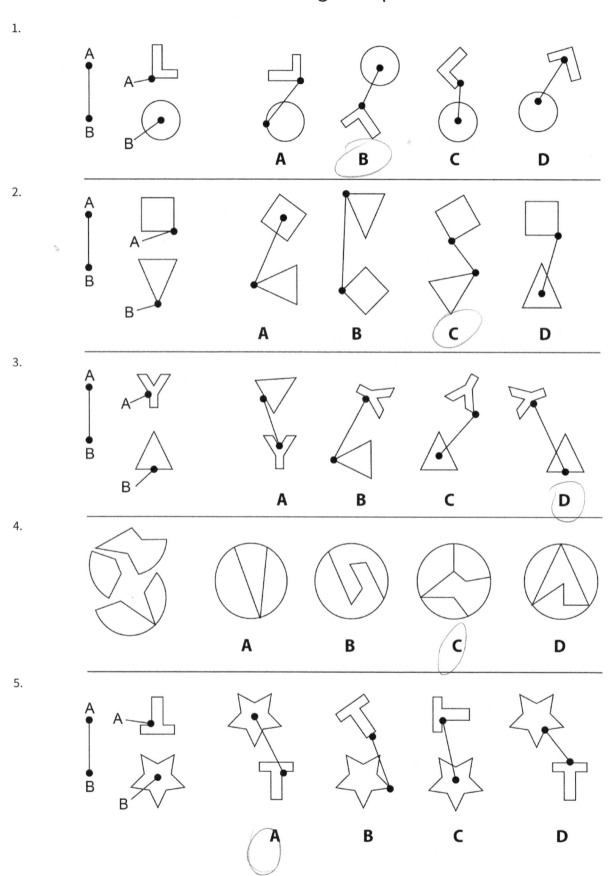

2.

3.

4.

5.

6.

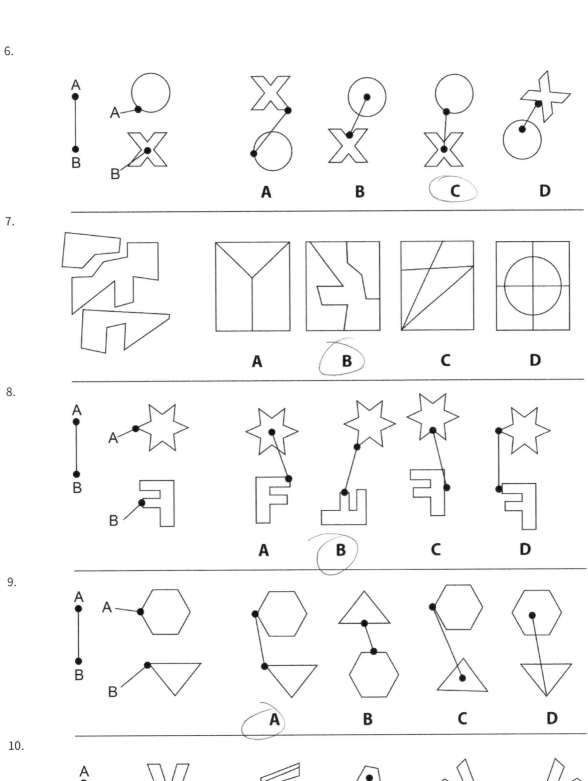

A   B   **C**   D

7.

A   **B**   C   D

8.

A   **B**   C   D

9.

A   B   C   D

10.

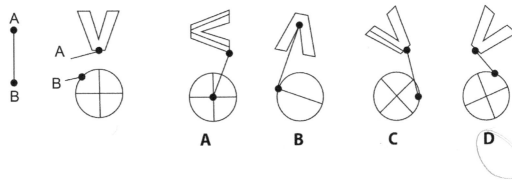

A   B   C   **D**

11.

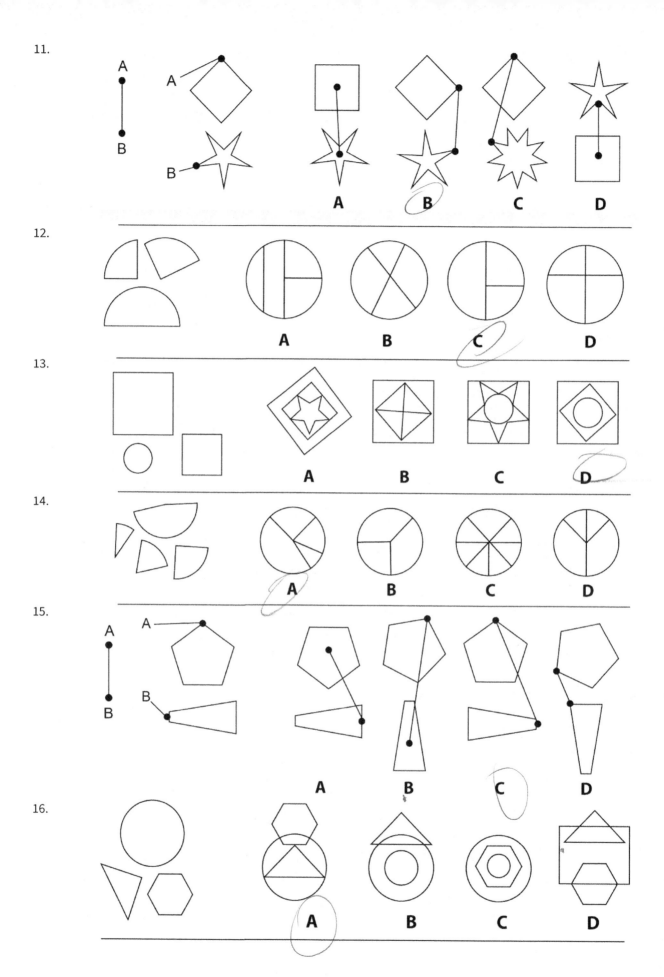

12.

13.

14.

15.

16.

Practice Test Four Answer Key

51/145

64.8%

GENERAL SCIENCE

1.	B.	5.	A.	9.	C.	13.	A.
2.	C.	6.	D.	10.	A.	14.	B.
3.	B.	7.	B.	11.	D.	15.	A.
4.	A.	8.	B.	12.	B.	16.	C.

ARITHMETIC REASONING

1.	B.	5.	A.	9.	C.	13.	D.
2.	D.	6.	C.	10.	B.	14.	C.
3.	D.	7.	B.	11.	B.	15.	B.
4.	D.	8.	A.	12.	C.	16.	C.

WORD KNOWLEDGE

1.	A.	5.	A.	9.	B.	13.	C.
2.	C.	6.	A.	10.	D.	14.	A.
3.	C.	7.	D.	11.	A.	15.	B.
4.	A.	8.	B.	12.	A.	16.	C.

PARAGRAPH COMPREHENSION

1.	D.	4.	A.	7.	A.	10.	C.
2.	B.	5.	B.	8.	C.	11.	B.
3.	C.	6.	C.	9.	A.		

MATHEMATICS KNOWLEDGE

1.	D.	5.	D.	9.	B.	13.	B.
2.	A.	6.	D.	10.	C.	14.	D.
3.	C.	7.	A.	11.	C.	15.	A.
4.	B.	8.	B.	12.	D.	16.	D.

ELECTRONICS INFORMATION

1.	B.	5.	B.	9.	B.	13.	C.
2.	A.	6.	D.	10.	C.	14.	D.
3.	C.	7.	C.	11.	D.	15.	B.
4.	D.	8.	B.	12.	A.	16.	A.

Auto and Shop Information

1.	D.	7.	C.	13.	C.	19.	B.
2.	A.	8.	C.	14.	C.	20.	B.
3.	D.	9.	D.	15.	A.	21.	D.
4.	B.	10.	A.	16.	C.	22.	D.
5.	B.	11.	C.	17.	A.		
6.	A.	12.	B.	18.	D.		

Mechanical Comprehension

1.	A.	5.	A.	9.	A.	13.	B.
2.	B.	6.	C.	10.	D.	14.	B.
3.	D.	7.	A.	11.	B.	15.	C.
4.	A.	8.	C.	12.	A.	16.	D.

Assembling Objects

1.	B.	5.	A.	9.	A.	13.	D.
2.	C.	6.	C.	10.	D.	14.	A.
3.	D.	7.	B.	11.	B.	15.	C.
4.	C.	8.	B.	12.	C.	16.	A.

seventeen

CONCLUSION

By reading through this guide, you should gain a good understanding, at the very least, of what is being tested and what is on the ASVAB. Again, your career in the military partially depends on how well you are able to do on this test. One thing you should remember is that fact that, though the ASVAB is scored and the scores are used, it is not the type of test that you can actually fail. These tests have been designed to test your knowledge. If you do not do as well as you want, then you can always retake the test. Don't let your recruiter try to railroad you through so he can get his pay off of you and meet a quota, take your time and move steady.

Retaking the test primarily depends on a number of situations, depending on the branch:

Army
The Army will only let you retake if you have failed to get a score high enough to enter service, an emergency or something occurred which prevented the completion of the test, or your ASVAB score has expired (two years after the date which you first took it). You cannot take a test to reach an enlistment bonus score and you cannot retake the test to get entry into a certain program or specialization.

National Guard
The National Guard works exactly in the same way that the army does. The National Guard will only let you retake if you have failed to get a score high enough to enter service, an emergency or something occurred which prevented the completion of the test, or your ASVAB score has expired (two years after the date which you first took it). You cannot

take a test to reach an enlistment bonus score and you cannot retake the test to get entry into a certain program or specialization.

Air Force

The Air Force has pretty strict rules. You cannot take the exam again if you are in a DEP (Delayed Entry Program). The test can be taken if your AFQT score is passing but you have not joined DEP yet. Retesting can also be allowed when the recruit wants to increase their line scores in order to get into a specific job to match qualifications.

Navy

The Navy will allow scores over two years old to be retaken and test retakes can also be taken if an individual fails to qualify for enlistment. Once you have been moved into DEP, the test cannot be retaken. An AFQT score between 28 and 30 will allow you into the DEP Enrichment Program, which trains you academically to allow you to retest and potentially increase your scores.

Marines

Tests can be retaken for the Marine Corps if the scores are two or more years old. If your recruiter deems it necessary or warranted, they can authorize a retest by request. It is important to note that you are not allowed to take a retest simply based on the fact that you did not get a high enough score for a particular job or specialization that you had your eye on. Otherwise, it is acceptable.

Coast Guard

The coast guard works exactly like the Marine Corps, but with one slight difference: You are allowed to retake the test to raise your scores for a certain specialization. The only catch to this is the fact that you will have to wait six months prior to taking the test again.

Finally...

Most of all, good luck with your test. Embarking on the journey to join the military is a commitment which should not be taken lightly. You get out of it what you put in. The better you are able to do on the ASVAB, the more options you will have and the more doors will be open to you. Take advantage of as much study time and as many materials as you possibly can as you prepare for that.

Made in the USA
Middletown, DE
08 February 2021